Transient Free Surface Flows in Building Drainage Systems

T0179065

Climate change will present a series of challenges to engineers concerned with the provision of both building internal appliance drainage networks and rainwater systems within the building boundary, generally identified as the connection to the sewer network. Climate change is now recognised as presenting both water shortage and enhanced rainfall design scenarios.

In response to predictions about imminent climate change *Transient Free Surface Flows in Building Drainage Systems* addresses problems such as the reduction in water available to remove waste from buildings, and conversely, the increase in frequency of tropical-type torrential rain. Starting with introductory chapters that explain the theories and principles of solid transport, free surface flows within drainage networks, and attenuating appliance discharge flows, this book allows readers from a variety of backgrounds to fully engage with this crucial subject matter. Later chapters apply these theories to the design of sanitary and rainwater systems. Case studies highlight the applicability of the method in assessing the appropriateness of design approaches.

In this unique book, research in modelling for free surface flows at Edinburgh's Heriot-Watt University is drawn on to provide a highly authoritative, physics-based study of this complex engineering issue.

John Swaffield was President of the Chartered Institution of Building Services Engineers for 2008–09, and Emeritus Professor of Building Engineering at Heriot-Watt University, Edinburgh, UK, until he tragically passed away in early 2011. The writing of this book was completed by his colleagues at Heriot-Watt University.

Michael Gormley is a Senior Lecturer in Architectural Engineering in the School of the Built Environment at Heriot-Watt University. He has been an active researcher in the field of fluid flow modelling since 2000. His research interests include pressure transient propagation and suppression in high rise buildings, water conservation and the modelling and control of infection spread in hospitals.

Grant Wright is a Lecturer in Civil Engineering in the School of the Built Environment at Heriot-Watt University. His research interests include fluid flow modelling at multiple scales, ranging from curtilage level drainage systems through to regional scale flood modelling, as well as the performance of sustainable urban drainage systems and public perception of flooding related issues.

Ian McDougall is a Computing Officer in the School of the Built Environment at Heriot-Watt University. He has been responsible for the production and maintenance of drainage-related computer models since 1995. His research specialisms are solid transport in horizontal drains and water conservation.

Transient Free Surface Flows in Building Drainage Systems

John Swaffield
with Michael Gormley, Grant Wright
and Ian McDougall

Routledge
Taylor & Francis Group

LONDON AND NEW YORK

First published 2015
by Routledge

2 Park Square, Milton Park, Abingdon, Oxfordshire OX14 4RN
52 Vanderbilt Avenue, New York, NY 10017

Routledge is an imprint of the Taylor & Francis Group, an informa business

First issued in paperback 2019

British Library Cataloguing-in-Publication Data
A catalogue record for this book is available from the British Library

Library of Congress Cataloging in Publication Data
Swaffield, J. A., 1943–
Transient free surface flows in building drainage systems/John
Swaffield, Michael Gormley, Grant Wright and Ian McDougall.
pages cm
Includes bibliographical references and index
1. Hydraulic transients. 2. Drainage. 3. Sewerage. 4. Runoff. 5. Drainage
pipes. I. Title.
TC171.S94 2015
696'.13—dc23
2014030683

ISBN: 978-0-415-58915-4 (hbk)
ISBN: 978-0-367-37780-9 (pbk)

Typeset in Sabon
by Swales & Willis Ltd, Exeter, Devon, UK

Contents

Illustrations

Tables

Foreword

My husband John died before this book was completed, and I am grateful that members of the Drainage Research Group (DREG) at Heriot-Watt University agreed to take on the task to complete and publish it. I am indebted to them.

John was always a 'research fellow' at heart and wanted to leave his research and books based on his life's work as a legacy to his new found knowledge. This has now been secured. But not only has he left his research to posterity – he has also left a team of researchers that have been inspired by his work and will take this forward in the future. I hope this book will also stimulate your interest in fluid mechanics applied to drainage systems as well.

Dr Jean Swaffield
February 2014
Edinburgh

Contributors' Foreword

When John Swaffield conceived this work it was intended to be a companion book to 'Transient Airflow in Building Drainage Systems', which dealt with building drainage ventilation and system design from an air flow and air pressure point of view. Taken together, these two books were to encapsulate John's work in building drainage research spanning over 40 years. It is safe to say that John Swaffield was the leading academic authority on building drainage system research, and these two books were intended as his legacy to future generations of researchers.

It was with great sadness that we undertook to complete this book after John's death in 2011; however, it was felt that this book needed to be finished in order to complete the pair.

We are in little doubt that this book would have been different had John finished it himself. His unique perspective and experience in this field of study would have given the conclusions a very personal flavour. Having said that, we have tried to be as faithful to his views as much as possible, drawing on his writings as much as possible, but also on personal conversations and correspondence with him.

John is sorely missed by all who worked with him. We hope that we have done his work justice in this book and that his legacy will continue to shape the work of other researchers in this field for years to come.

Michael Gormley
Grant Wright
Ian McDougall
Edinburgh
September 2014

1 Water is the new carbon

However it is viewed, the movement of water forms a considerable component in the taming of the environment in our modern world. From the channelling of flows for water supply to the transport of waste, sewerage and rainwater away from habitable space, this important aspect of engineering is both taken for granted and ubiquitous in equal measure. The challenge has always been to understand a system that most users are unaware of. It could be argued that the current challenge of 'climate change' occurs on a continuum, from a developed world perspective, encapsulating issues such as social reform and the linkage of sanitation systems to public health issues in the nineteenth century, large infrastructure projects post both world wars in the twentieth century, environmental concerns in the 1970s and 1980s and continuing water conservation and resource management concerns up to the present. From the global perspective it can be noted that the rapid urbanisation in the last half of the twentieth century leaves more than half the world's population without clean water supply and adequate sanitation provision (Gormley *et al.* 2013), arguably, at a different point in the continuum. While the nature of the challenge changes, the fundamental engineering questions remain very similar.

From a practical perspective, the recent focus on a changing climate in relation to water and wastewater removal from buildings centres on the ability of systems to continue to operate efficiently with the changing demands placed on their functionality from the extremes of too much water and too little water. What has become clear is that the issue of climate change has become more than a discussion of carbon emissions, but rather a wider discussion on mitigation and adaptation approaches which inevitably lead to the need for more holistic approaches. As a fundamental of human existence, water availability and management is seen as central to the debate, leading some to contend that 'Water is the new Carbon' (Swaffield 2008).

It has been established, and widely accepted, that climate change has led to a multiplicity of water-based problems, ranging from the accepted increase in sea levels to increased hurricane frequency and severity and precipitation increases due to rising sea water temperatures. Along with the probability of increased severity and frequency of rainfall, with consequent flooding, water

shortages are also a challenge for systems engineers. Areas of water stress in
the UK have been established (Figure 1.1), and similar issues are raised in
other areas.

UK government predictions for climate change, June 2009, aimed at indus-
tries and organisations that need to make long-term investment decisions
that could be influenced by climate change, suggest that summer rainfall in
SE England could fall by 19% by 2050 and possibly (1 in 10 chance) by
41%. Winter rainfall in the West of Scotland will rise by 15% by 2050 and
possibly (1 in 10 chance) by 29%. In launching these new scenarios Hilary
Benn, the UK Environment Secretary said that 'climate change is the biggest
challenge facing the world today . . . this landmark scientific evidence shows
that we need to tackle the causes of climate change and deal with its consequences' (Guardian report of Benn's presentation, 19 June 2009).

In the UK, the importance of building water supply and drainage has
been promoted by government departments such as the Department of
Communities and Local Government (DCLG) and the UK Department for
Environment, Food and Rural Affairs (Defra) in their development of new
policies and advice to designers and users in the Part G Building Regulations,
and a companion publication, in the CIBSE Guide series, *CIBSE Guide G:
Public Health Engineering* (Chartered Institution of Building Services Engineers
2009). Similar policies have been followed by governments around the world.
Handling the consequences of climate change requires research-led initiatives
that feed through to dissemination of subject knowledge and design guidance.

The water-based climate change issues shown in Figure 1.2 have all led to
policy-related initiatives, and while this sets political agendas, the engineering
fraternity have two major advantages in dealing with climate change – there
is the intellectual ability to understand the science of our planet and the

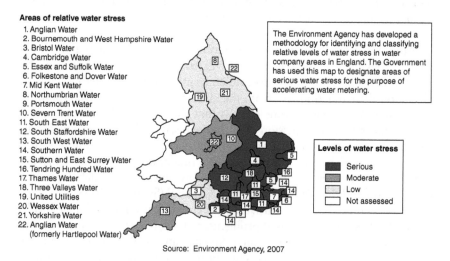

Source: Environment Agency, 2007

Figure 1.1 Levels of relative water stress in the UK.

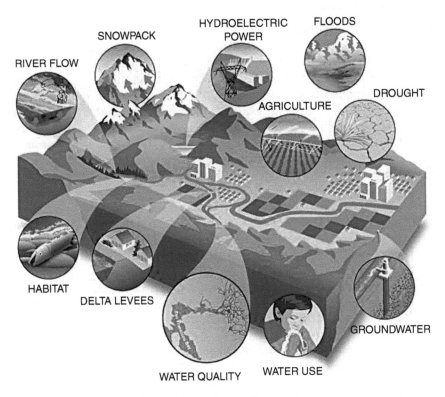

Figure 1.2 Water-based climate change issues. (Source: Environment Agency).

innovative ability to transform that scientific understanding into engineering solutions. Efforts to mitigate climate change will require more rather than less technology to stand beside the renewable 'green' agenda. Therefore it is expected that 'big engineering' in the form of tidal barrages, new hydroelectric proposals, an enhanced nuclear industry, the development and application of carbon sequestration and a possible national water grid will be solutions that should be welcomed, possibly providing both climate change modification as well as Rooseveltian economic solutions (Swaffield 2009b).

1.1 Too much or too little?

The challenges associated with the provision and management of water in the built environment centres around two major contradicting phenomenon: too much and too little water. This is clearly an over simplification of many of the complex water supply and drainage management issues facing professionals in the industry; however it is a useful starting point in classifying the challenges.

1.1.1 Too little water

Globally, the distribution of water availability is uneven, dependant on local resources, both natural and financial, leading to a water stress. Water stress is defined by the intergovernmental panel on climate change (IPCC) as a situation in which the availability of water 'directly affects human activity' and has been quantified as an availability of 1000 m³/capita/year. In many parts of the world (in arid areas for example) there is a considerable requirement from systems which have become or which are becoming depleted. Work by Arnell (2004) highlights the expected changes in water stress around the world under a range of climate change scenarios. Using a number of respected climate change models (HadCM3, ECHAM4, CGCM4, CSIRO, CFDL and CCSR) the work concluded that by the 2080s, with the exception of South Asia and the Northwest Pacific and some parts of West Africa, more people will experience an increase in water stress than will experience a decrease. The case for North America, Europe and the Mashriq[1] is expected to be particularly severe as is central and Southern America, although to a lesser extent.

It can be argued that the change in water supply per capita requirements creates a challenge in itself for engineers seeking security of supply and quality. A cursory look at the per capita water usage in different places highlights the nature of the problem.

Daily per capita use of water in residential areas:

- 350 litres in North America and Japan
- 200 litres in Europe
- 10–20 litres in Sub-Saharan Africa.

Efforts are inevitably focussed on reducing the daily per capita usage for North America, Japan and Europe while trying to increase the daily per capita in Sub-Saharan Africa. The imbalance in population, and population growth, coupled with insufficient infrastructure and water resources makes this a monumental task.

As the resource is becoming scarce, tensions among different users may intensify, both at the national and international level. Table 1.1 shows a classification of water stress levels on a global scale, and while developed countries are not suffering in the same way at present, higher levels of stress are expected as resources become more scarce. Over 260 river basins are shared by two or more countries. In the absence of strong institutions and agreements, changes within a basin can lead to trans-boundary tensions. When major projects proceed without regional collaboration, they can become a point of conflict, heightening regional instability. The Parana La Plata, the Aral Sea, the Jordan and the Danube may serve as examples. Due to the pressure on the Aral Sea, half of its superficy has disappeared, representing two-thirds of its volume. 36 000 km² of marine grounds are now covered by salt.

Table 1.1 Water resources index classes

Resources per capita (m³/capita/year)	
Index	Class
>1700	No stress
1000–1700	Moderate stress
500–1000	High stress
<500	Extreme stress

Source: Arnell 2004.

UN initiatives have consistently failed to achieve their objectives due to the scale of the problem, lack of resource and motivation and political instability in areas most in need of support. The 1980s UN 'Decade of water' singularly failed to meet its objectives, and the necessity to improve developing country water and sanitation provision, together with the challenges posed by global urbanisation, continues to be daunting. The current millennium development goals (MDGs) will have fallen short of the sanitation target by a long way, while access to water supply is improving (UN 2011).

In the UK, along with other developed countries, water shortage is exacerbated both by local shortfalls in precipitation and by our choices as to urban location and life style. Current Defra proposals indicate a reduction in water usage from 150 litres per capita per day to 120–130 by 2030. This aspiration has to be seen against a background of climate change that over the same period is expected to result, in the southern UK, in high summer temperatures and dry conditions, while extreme winter precipitation will become a normal event.

While efforts have inevitably focussed on water for drinking, since this is a fundamental requirement for a healthy human existence, there are water-related issues from a sanitation perspective.

The issue is often divided between engineering challenges in the 'developed world' and the 'underdeveloped' world making a distinction based on infrastructure and resources.

Table 1.2 highlights the global challenges for water supply and sanitation, regardless of economics. There are differing and often competing demands on systems; for example, systems must have enough water to be self-cleansing whilst not overusing valuable potable water. Alternatives are complex and often counter-intuitive. For example, in order to reduce the risk of blockages in a system with low water use, it is often prudent to decrease the pipe diameter, thus increasing the water depth, decreasing the wave attenuation and facilitating better solid transport.

A cursory glance at Figure 1.3 indicates the extent of the challenges. In 1900 the average flush volume was 40 litres; today, with even further drives hoped for the pressure is to reduce on 6 litres, with 4/2 litre flushes not uncommon. It is easy to imagine the different solid transport characteristics with such a difference in applied surge.

Table 1.2 Global water challenges

	Developed world	Underdeveloped world	Developing world
Water resource for drinking	Generally available – high cost and subject to localised stress Water usage per capita high Changing demographics	Availability variable Lack of infrastructure	Rapid growth in distribution and processing infrastructure difficult to manage (e.g. Brazil, India and China)
Water resources for sanitation	Drive to reduce high levels Concerns over ability of current infrastructure to cope Potable water usage high – opportunities for recycling at relatively high cost to user. Changing demographics	Generally none used. Exceptions are small-bore simplified sewerage. Low-cost waterless systems still favoured by government and NGOs.	Tension between on-site ultra-low, or zero water usage systems and high-water-usage conventional sewerage systems.

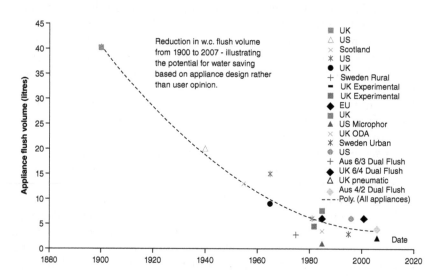

Figure 1.3 Reductions in w.c. flush volume since 1880.

Careful drainage system and appliance design can minimise the probability of deposition while flow booster solutions should also be recognised. Minimum transport distance to the first joining flow junction is an essential parameter and has design implications, as shown in Figure 1.4. Some important water closet (w.c.) design parameters are shown in Figure 1.5 also.

It can be seen that there are a considerable number of variables with which to make improvements, and it remains an imperative of the industry at large to continue to strive for improvement.

Similarly junction design becomes a major issue as the hydraulic jumps upstream of a junction of two or more flows present an impediment to solid transport leading to deposition. Swept entry junctions should be used and top entry 90° entries banned.

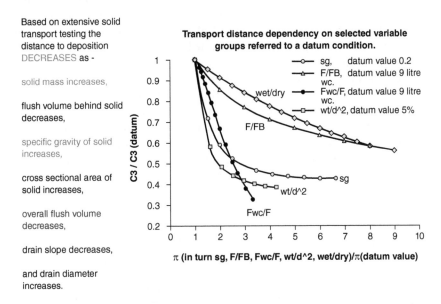

Based on extensive solid transport testing the distance to deposition DECREASES as -

solid mass increases,

flush volume behind solid decreases,

specific gravity of solid increases,

cross sectional area of solid increases,

overall flush volume decreases,

drain slope decreases,

and drain diameter increases.

Figure 1.4 Solid transport dependence on w.c. design parameters, as well as both drain and waste solid dimensions.

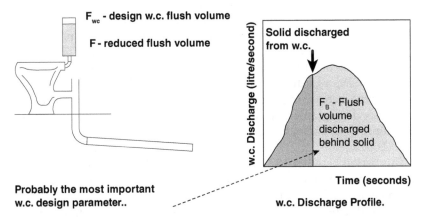

Figure 1.5 Solid transport depends primarily on the volume of flush water discharged behind the solid. Maximising this volume is a major design objective.

Innovation has not always been smooth, nor have lessons necessarily been learned. During the 1970s Scandinavia was regarded as leading the field in efficient appliance design; however, undeniably due to the extended droughts suffered in the past decade, Australian industry has now taken on this mantle. Despite w.c. water usage reducing internationally, a time traveller present in 1900 at the discussions between the Metropolitan Water Board and the appliance manufacturers aimed at reducing flush volume from 40 litres, who then moved forward to London in 1998 to attend meetings of Defra's Water Regulations Advisory Committee, would have been amused at two things that would not have changed over the 100 years – the speed of travel across the capital and the arguments put forward by the manufacturers to resist reductions in flush volume.

Flow boosting is also of interest. Research in the 1980s concentrated on two alternate designs: the traditional tipping tank used in the UK from at least the 1860s, where an eccentrically pivoted tank tips a large water volume into the drain at periodic intervals, and the siphon tank, accepted by Stockholm in 1989 as a design solution to the installation of 3 litre w.c.'s in city apartment blocks – a 21 litre siphon tank in the basement intercepted w.c. flushes and delivered its contents to the drain in one surge.

In addition to the technicalities of water availability and an ability to model system flows there are 'organisational' issues to contend with. Water supply and drainage for buildings relates to randomly occurring events in a system, principally operated by users, with their own requirements for hygiene and standards of cleanliness. In addition to users' preferences there is the issue of capacity, or more correctly in this case, loading. There is one such loading phenomenon which has come to the fore recently. It is relatively specific to developed countries; however it is widespread among them. A rapidly changing demographic which has led to an increase in single occupancy dwellings presents a particularly peculiar water supply challenge. Figure 1.6 indicates the issue well. The relationship between the number of occupants in a property and water consumption per capita is not linear. For example a single occupancy house may use approximately 200 litres per day; if a house has six occupants, then the total is 600 litres per day (100 litres/person/day). It appears that there is a 'baseload' requirement for a house to function which is needed regardless of number of occupants. There is a possibility that the apparent increase in per capita use per day is in fact due to the changing occupancy demographic. An increase in usage of 35% in UK average daily consumption per capita between 1971 to 2001 has been estimated by Moran *et al.* (2007).

This period between 1971 and 2001 saw rapid changes in lifestyle in the developed world, and the trend towards more single occupancy continues unabated, as can be seen by Figure 1.7.

The need to be able to assess these complex and counter-intuitive arrangements lends itself to a modelling technique in which the vast range of possibilities can be predictive and analysed. This book seeks to set out the

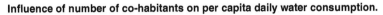

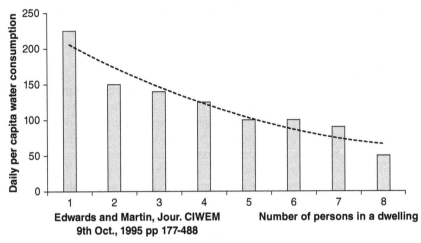

Figure 1.6 Influence of number of co-habitants on per capita water consumption.

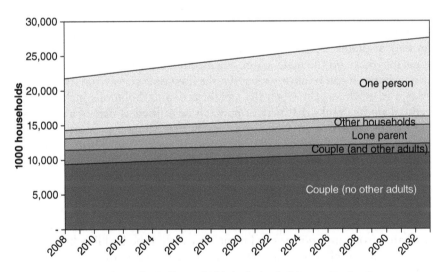

Figure 1.7 Projected number of households by household type, England.

breadth and depth of work carried out in this area over the past 50 years or so and presents a numerical model, DRAINET, capable of providing the engineer and policy maker with the best information possible to drive legislation and regulation. A bonus comes from the true understanding of system operation with the possibility of proposing novel approaches and devices which will inevitably be needed globally to tackle rapid changes in water demand and supply. The mechanisms involved in assessing solid transport through mathematical modelling are discussed in Chapter 4, while the techniques employed to maximise system performance are discussed in Chapter 5 of this book.

1.1.2 Too much water

Principally the issue of too much water relates to increases in precipitation and the ability of systems to cope with removal from the built environment. The area of concern here is roof drainage and the particular methods associated with design and performance analysis. Modelling roof drainage systems is covered in Chapter 6, and the context for that research is given here. An example of a large airport installation is shown in Figure 1.8, highlighting the extent of the design and the low number of downpipes required for such an expansive roof.

While there are considerable challenges relating to a variability in rainfall, there are opportunities as well. The availability of 'un-processed' water can be harvested and stored to provide additional water for tasks such as irrigation and w.c. flushing; both are heavy users of water and do not require potable water.

The introduction of siphonic roof drainage systems in the 1960s revolutionised the way roofs were drained. Figure 1.9 indicates the different surge profiles available from conventional and siphonic systems and the effect these may have on the main sewer when combined with other outflows in a storage tank. It can be clearly seen that the flow/time curve from the siphonic system is much steeper when the siphon activates. Solutions include intervention storage tanks that may or may not be utilised as part of a rainwater harvesting system or so-called green roofs that allow a proportion of the precipitation to be absorbed or re-evaporated to the atmosphere, or indeed, a reversal of the popular urban replacement of front gardens with vehicle parking.

Current predictions indicate that UK rainfall severity and frequency will increase during the winter months. To avoid surcharging existing sewers we must understand the techniques necessary to attenuate and delay the peak flow discharge from any roof or catchment area within the building boundary – basically an application of the unsteady continuity of flow equation, where inflow equates to outflow minus the rate of change of storage within the control volume – in this case the building or property boundary. This requires an understanding of the time-dependant nature of rainfall-induced sewer flows.

1.2 Grey water reuse

The reuse of 'grey' water represents a useful way to reduce the daily water demand for a building. In many applications the largest savings in mains water are likely to be obtained by using reclaimed water for toilet flushing

Stanstead Airport
Roof area: 39,204 m²
Roof outlets: 132
Downpipes: 16

Figure 1.8 Siphonic system on Stanstead Airport.

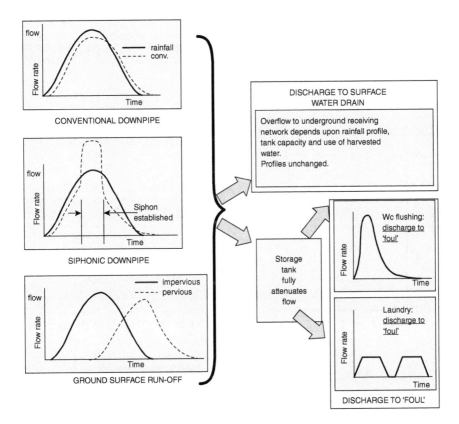

Figure 1.9 Siphonic system unsteady inflow/storage/outflow.

(Pennycook *et al.* 2007). Low public acceptance of using grey water for activities such as watering vegetables has been widespread, and it could be suggested that users may prefer to use rainwater for such activities and use grey water for non-personal activities such as toilet flushing. This would certainly improve the acceptability of water recycling systems, especially since it has been suggested that public acceptability improves after exposure to such systems (Hills *et al.* 2002). It would also fit in with the views of Dolnicar and Schafer (2009) who want to exploit the powers of word of mouth and use influential people to endorse and publicise these alternative systems.

It has been suggested that there is a 'cumulative flow balance' between the grey water collected and the volume of water required for the w.c.'s (Jefferson *et al.* 2000; Diaper *et al.* 2001). Jefferson further explains that, although there is this 'cumulative flow balance', grey water is generated over short time periods and not always in tandem with toilet flushing, which occurs more consistently throughout the day. Figure 1.10 depicts these variations in times of supply and demand, that will generally result in a deficit in water during the afternoon and later evening (Surendran and Wheatley 1998), which therefore require the recycled water to be stored to balance out the variations between generation and use. However, it should be noted

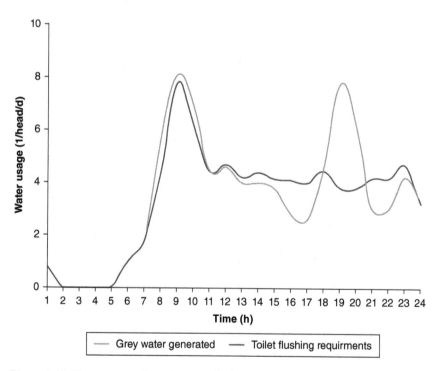

Figure 1.10 Grey water collected v. w.c. flushing requirements. (Source: Jefferson *et al.* 2000, courtesy of Surendran and Wheatley 1998).

that residence time in systems dramatically affects the characteristics of grey water, and care should be taken to ensure that grey water is not stored for long periods of time. An investigation by Dixon *et al.* into storage tanks found that a 1 m³ tank was suitable for a wide range of occupancy scales (Dixon *et al.* 2000). Jefferson *et al.* found that increases in storage capacity over 1 m³ provided marginal rises in water saving whilst also enhancing problems associated with grey water degradation and disinfection reliability, due to prolonged storage (Jefferson *et al.* 2000; Dulcinar and Schafer 2000).

While grey water recycling provides a 'positive' on the water conservations side, it is not without its negative attributes too. One of the consequences of recycling grey water is that this represents a reduction in the availability of water to keep systems clean. As flush volume decreases, there is an attendant reduction in solid transport distance, in many cases systems rely on adjoining flows from other appliances to keep the system clear. The problem is illustrated in Figure 1.11 (Gormley and Dickenson 2008).

The cleansing effect of the long bath discharge is considerable in relation to the small duration w.c. wave. The effects of removing all large volume discharges from a system serving a small housing estate with 10 houses was investigated using the numerical model DRAINET. The conclusion of the study revealed that reducing flows from grey water in this context was acceptable; however it did highlight that care needs to be taken with the arrangement of adjoining branches and flows from other appliances. Figure 1.12 shows the critical distances developed from the study. This example shows a very practical approach to the issue of reducing flows as a consequence of a measure used to mitigate against climate change (grey water recycling). The study also

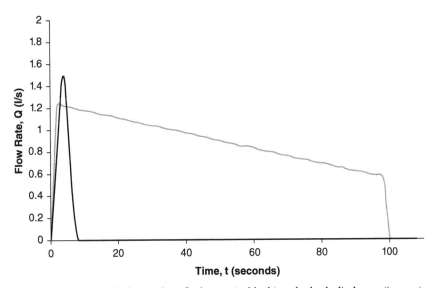

Figure 1.11 Q/t graphs for a 6 litre flush w.c. (in black) and a bath discharge (in grey).

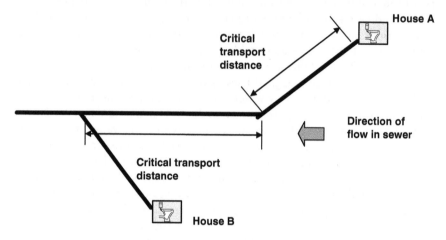

Figure 1.12 Critical solid transport distances for multi-house installation.

shows that for many 'real world' applications, the story lies in the nuances of the problem; in this case it is not 'maximum transport distance' that is important, but the distance to the next adjoining flow. This example highlights the power of a robust numerical model combined with experience of how real systems work and having an open-minded approach to a problem involving complex interacting systems. This method was expanded and used to inform the new 'Sustainability' section of the Scottish building standards (Gormley 2011) which embedded reductions in carbon emissions in the whole building design and implementation process.

1.3 Modelling a new future

For any solution to have realistic capabilities requires an in-depth understanding of not only the functionality of specific techniques and technological solutions but also their interaction with existing systems. It is not uncommon for a new technology to work perfectly well on its own, only to render some other essential component redundant or ineffective (Swaffield 2009a).

It can be seen from the grey water recycling example in the previous section that numerical modelling offers an additional insight which is not readily reproducible in a laboratory and is difficult to measure on site. In terms of the insights which can be gained, the central importance of w.c. flush volume in the water conservation debate is still very important, as are the efforts to reduce the water usage attributed to w.c. operation internationally. The fundamental prerequisites of low flush w.c. design will be revisited, including the importance of the percentage of the flush discharged behind the solid. Similarly the strategies available to enhance solid transport are important, as are both the simplistic 'reduce drain cross sectional area and increase drain slope' options and the more sophisticated solutions based on the introduction of flow boost appliances.

An understanding of this sort is greatly enhanced by the extent of research and investigation required in order to effectively model whole systems. In addition to the knowledge gained in such research, the output for addressing these issues requires a design capability that recognises the necessity to deal with the interpretation of data that will indicate the changing patterns, frequency and severity of rainfall.

Research requires an interaction between industry (taken here to include design consultants, contractors, building specifiers and operators), government (taken to include both local and national and edicts emanating from trans-governmental organisations) and dedicated research providers (taken here to include both academia and research institutes, whether governmental or industrial).

The effects of climate change within the built environment include the following:

- Increased frequency and severity of rainfall
- Water shortages and the introduction of water conservation strategies, to include solid transport.

Modelling solid transport also requires simulation techniques to represent both the attenuation of an appliance discharge during its passage along a branch drain and the interaction between solids in transport and the surrounding flow. These models are at their most challenging in dealing with the attenuating free surface flow waves generated by appliance discharge that are only found in the upper reaches of the drainage network. Models dealing with the transport of solids in the quasi-steady flow encountered in sewer systems downstream of the building drain connection are of necessity somewhat simplistic by comparison, as the rate of change of accompanying flow will be low and the flow depth surrounding the solid will be greater than in the building network. Method of characteristics simulations to model wave attenuation in free surface unsteady flow were developed to allow the addition of solids in transport as moving boundary conditions within the branch drain (McDougall 1995; McDougall and Swaffield 2003).

In order to provide a basis for the future prediction of waste solid transport it is useful to first review the fundamental research that developed the current modelling capability. In addition it is important to identify the deformable solid parameters that contribute to the transport performance encountered in any drainage system. It will also be necessary to compare the early experimental research with later but related simulations of solid transport to both corroborate the descriptions of solid transport and provide a basis for prediction that may be used to inform system design.

The fundamentals of fluid flow in open channels or partially filled pipes underpins the theory behind free surface wave modelling in building drainage systems. The solution of the St. Venant equations of momentum and continuity wholly describes wave propagation and attenuation in system pipeworks, and the Method of Characteristics provides a calculation framework from which the calculation of initial conditions and boundary conditions can be allowed to progress. Together, these techniques, backed up by over 40 years of laboratory investigation and site based studies have been used to produce a suite of free surface wave models – namely DRAINET, FM4GUTT, ROOFNET and SIPHONET. The bases of these powerful models are described in the chapters that follow, and have the capability of modelling entire system operation in an accurate, repeatable and robust manner.

1.4 Summary

Climate change in the present is real and offers both challenges and opportunities for engineers and academics in the future. The modelling of free surface water flows using a Method of Characteristics approach in a finite difference scheme has been made possible by increased availability of computing power in the last 30 years. The solution of the St. Venant equations of momentum and continuity is at the heart of the technique and has proven to be reliable, insightful and robust. Run times have reduced to the point where flows in complex systems can be modelled in a matter of minutes. The 1-D flow model (including some 2-D models) provides sufficient data to analyse flow phenomenon and test configuration hypotheses. This 'look see' approach has been useful in advising government (Part H regulations and Scottish Section 7 regulations) and industry alike.

It is noteworthy to mention that the current modelling capabilities are still not feasible using computational fluid dynamics (CFD) methods. Unless supercomputing power is available the run times for complex systems is still weeks and months. There are methods in development – smoothed particle hydrodynamics (SPH), for example – which promise increases in speed; however these require the solution of the *Lattice Boltzmann* equations in favour of the *Navier–Stokes* equations which is computationally challenging. For now, and until alternative mesh-less solutions become more widespread, it is still preferable to use the Method of Characteristics models described in this book for large pipe networks.

While the fundamental St. Venant equations effectively deal with wave attenuation in near horizontal pipes, practical application of the model requires a set of boundary conditions to root the model in the physical world. A set of equations describing physical phenomena such as w.c. discharge surge, pipe junctions and inlet conditions, solid movement and deposition, hydraulic jump phenomena, siphonic rainwater outlet flow conditions, gutter flow, siphonic balancing (including the MacCormack/MoC hybrid model) make this suite of free surface flow models extensive in the context of modelling changing flow conditions in a building.

This book seeks to present a definitive exposition and analysis of free surface modelling relating to buildings, starting with the theory of fluid flow in open channels or partially filled pipes in Chapter 2, in which fundamental equations for flow and frictional representation are provided. Chapter 3 defines the solution of the St. Venant equations of momentum and continuity in relation to unsteady flows and their application in the Method of Characteristics numerical technique including an extensive discussion on boundary condition derivation and application. Chapter 4 introduces the mathematical techniques described in the previous two chapters specific to partially filled pipe flow in building drainage systems. One of the continuing strengths of the techniques described in this book is the ability to assess the performance of a building drainage system to perform its fundamental requirement – to remove potentially harmful, pathogen-laden waste from habitable space efficiently. In order to make this performance assessment, it is necessary to understand the mechanisms underpinning solid transport in unsteady flows in partially filled pipes. Chapter 5 introduces the concept of solid transport, as well as the work carried out historically on the topic, and provides the basis for solid transport boundary conditions for use in the numerical model. Roof drainage can be considered to be on the 'front line' in a world with a changing climate. Designs predominantly cater for prevailing weather conditions in a locale, and when that alters radically, problems may emerge with overtopping as rainfall intensity increases. Chapter 6 details free surface modelling of both conventional and siphonic rainwater drainage systems, including innovative hybrid Method of Characteristics/MacCormack method approach used for transition from free surface modelling to full-bore modelling as is required by siphonic systems. Chapter 7 provides practical examples of the applications of these methods while Chapter 8 draws the work together and concludes with predictions on how this body of work might move forward in the future.

This book provides all the theory, tools and techniques necessary to fully predict any eventuality in a building drainage system. Since the models described in this book are based on fundamentally sound laws of physics and tried and tested flow fluid engineering, backed up by extensive laboratory investigations, they are wholly appropriate for application to any building drainage system under any set of changing criteria, including climate change. The strength therefore in this approach is that it is fundamental and applicable both in a predictive sense – what happens if? – and also in a forensic manner – why did that happen? The benefits of being able to answer such questions are obvious and provide support and insight for designers, manufacturers, regulators and policy makers alike.

Note

1 The Mashriq region includes Iraq, Jordan, Lebanon, Syria and the occupied Palestinian territories.

2 Fluid flow conditions in open channels and partially filled pipes

The passage of water through building drainage and rainwater drainage networks may be characterised as predominantly free surface in nature, with the exception of surcharge conditions and siphonic action rainwater systems.

Free surface flow conditions are complicated relative to full bore pipe flows by the obvious fact that, while the air pressure above the fluid surface remains sensibly constant, the depth of flow, and hence the cross-sectional configuration of the filled conduit, will change. (In full bore flow the conduit cross section remains constant but the fluid pressure responds to flow changes). This effect is present in steady non-uniform free surface flow – defined as conditions at any section remaining constant with time but possibly varying with longitudinal position – as the flow depth responds to any local obstruction or junction by generating a backwater profile where flow depth changes in the flow direction under steady conditions. Similarly flow depth will change if the flow discharged to the conduit changes with time – this results in a wave effect that propagates in the flow direction or, in the event that the wave reaches a junction, in the reverse flow direction in some of the joining conduits.

The essential element of free surface flow to be understood in the context of building drainage systems is the mechanism necessary to transmit changes in flow condition within the flow, emanating from some change in boundary condition interaction with the flow. In full bore flow condition change is transmitted by the propagation of pressure transients whose magnitude depends on the fluid transported, the pipe cross section, its elasticity and the rate of change of flow velocity. Small changes to the conduit cross section may also accompany such transients. The velocity of such transients in the low-amplitude air pressure transient cases met in building drainage systems will be close to the acoustic velocity in air – some 320 m/s. In free surface flows any change in flow or boundary condition is communicated by the propagation of surface waves, and here the surface wave velocity is dependent upon flow depth. For a rectangular section the wave speed is $\sqrt{gh}$, where g is the acceleration due to gravity and h is the flow depth, a relationship yielding a wave speed of a 1 m/s for a 100 mm deep flow. Thus many flow conditions will have mean velocities that exceed the local wave speed so that flow disturbances cannot be transmitted upstream. Such flows are termed supercritical and are analogous to supersonic flows. Flow conditions where the mean flow velocity is less than the local wave speed are termed subcritical. Transitions from subcritical to supercritical flows occur at hydraulic jumps – equivalent to shock waves.

2.1 Fundamental equations defining free surface flows

It will therefore be helpful to define the terminology describing the fluid flow conditions under free surface conditions and then to develop the defining equations necessary to lay the foundation for a discussion of free surface flow mechanisms within building drainage networks.

Steady flow	Flow where the flow conditions at any position remain constant with time. Steady flow will occur in long channels with a uniform cross section set at a fixed slope. This condition balances the potential energy provided by the channel downwards slope with the frictional resistance to the flow. Under these conditions the flow depth is referred to as Normal depth. Note that close to the entry and exit from such channels there will be zones of steady non-uniform flow, also referred to as varied flow, where the flow depth changes in the flow direction.
Unsteady flow	Flow where the flow conditions at any position change with time.
Uniform flow	Flow where the flow conditions do not change with distance along the flow.
Non-uniform flow	Flow where the flow conditions do change with distance along the flow.
Subcritical flow	Flow where the mean velocity is less than the wave speed – characterised by deep, slow flows.
Supercritical flow	Flow where the mean velocity exceeds the local wave speed – characterised by shallow, fast flows.
Varied flow	Alternative terminology for steady non-uniform flows.
Critical flow depth	Flow conditions where the wave speed equals the flow mean velocity, a Froude Number of unity. Occurs whenever flow changes from subcritical to supercritical, for example immediately downstream of any flow obstruction of junction that has raised the upstream flow depth.
Hydraulic jump	A flow discontinuity analogous to a shock wave that allows supercritical flow to translate into deeper, slower subcritical flow. Found upstream of any obstruction or junction.

Figure 2.1 illustrates these flow conditions for a simple conduit terminated in an obstruction or a junction with other flows.

Consider a disturbance in the flow transmitted as a surface wave at velocity c and height δh, in an uniform but non-rectangular channel and constant slope, Figure 2.2, to represent the range of channel and partially filled pipe cross sections to be considered.

If this wave is brought to rest by the imposition of a velocity −c onto the system, the flow condition will appear steady to an observer, and the equation of continuity may be applied across the wave front:

$$A_1(u_1 - c) = A_2(u_2 - c) \tag{2.1}$$

as density is constant.

The equation of motion applied across the wavefront equates the out of balance hydrostatic forces across the wavefront, due to the depth change δh across the wave, to the rate of change of momentum across the wave, hence

$$\rho g(A_1 \bar{h}_1 - A_2 \bar{h}_2) = \rho(A_2(u_2 - c)^2 - A_1(u_1 - c)^2)$$

where $\bar{h}_{1,2}$ are the centroid depths upstream and downstream of the wave.

Substituting for $(u_2 - c)$ and simplifying yields

$$\rho(u_1 - c)^2 = g \frac{A_2}{A_1} \rho \frac{(A_1 \bar{h}_1 - A_2 \bar{h}_2)}{A_1 - A_2}$$

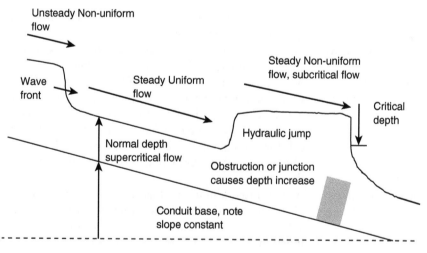

Figure 2.1 Schematic definition of free surface flow descriptors.

The net hydrostatic force, F_h, in the flow direction is effectively due to the increased hydrostatic pressure due to the wave height applied across the whole upstream flow cross-sectional area as follows:

$$F_h = \rho g(A_1\bar{h}_1 - A_2\bar{h}_2) = \rho g(\int_0^{h1_1} thdh - \int_0^{h2_1} thdh)$$

Splitting the hydrostatic force acting downstream into components due to h_1 and the wave height dh yields

$$\rho g(\int_0^{h1_1} thdh + \int_0^{h1_1} t(h + \Delta h)dh - \int_0^{h2_1} thdh) = \rho g(\int_0^{h1_1} t\Delta hdh - dA\bar{\Delta h})$$

where $\bar{\Delta h}$ is the centroid depth for the constant wave height Δh and

$$\int_0^{h1_1} t\Delta hdh = \Delta hA_1$$

Neglecting second-order terms reduces the downstream force expression to

$$F_h = \rho g\Delta hA_1$$

(Alternatively referring to Figure 2.2, it will be seen that the upstream and downstream hydrostatic forces due to depth h_1 cancel out, leaving the downstream acceleration force as the force due to the hydrostatic pressure at depth dh, i.e. ρgdh, acting over the area of the downstream flow plus the hydrostatic force determined for the wave height dh as $dA\bar{\Delta h}$. As A_1 tends to A_2 and as second-order terms, i.e. $dA\bar{\Delta h}$, may be neglected, the force may be expressed as $F_h = \rho g\Delta hA_1$).

As A_2 tends to A_1, A_2 / A_1 tends to unity, and as $A_2 - A_1 = dA = T\Delta h$, it follows from the momentum equation that

$$\rho(u_1 - c)^2 = \rho g\frac{A\Delta h}{T\Delta h} = \rho g\frac{A}{T}$$

Hence if the wave height Δh is small compared to h, then the velocity of propagation of the wavefront relative to the fluid is given, for any uniform cross section channel, by

$$u - c = \sqrt{\frac{gA}{T}}$$

$$(2.2)$$

Referring to Figure 2.2, the wave propagation velocity imposed on the system was negative; hence, as velocities downstream are positive,

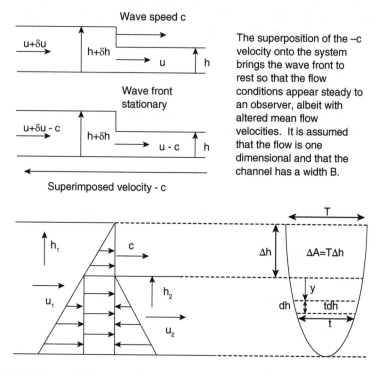

Figure 2.2 The superposition of the –c wave speed upon the system brings the wave-front to rest and allows the determination of the surface wave speed.

the propagation velocity of the wavefront downstream is $u + \sqrt{gA/T}$ and $u - \sqrt{gA/T}$ upstream. Hence if $u > \sqrt{gA/T}$, it follows that a wave cannot be transmitted upstream. The ratio of the flow velocity u to the velocity of propagation of the wave $c - u$ is known as the Froude Number, Fr, that defines much of free surface flow theory.

$$Fr = u / \sqrt{\frac{gA}{T}}$$ (2.3)

The Froude number defines the type of flow present in an open channel. If Fr < 1, then the flow is subcritical and disturbances may propagate upstream, while if Fr > 1, then the flow is supercritical and disturbances cannot be transmitted upstream – there is a obvious analogy here to the Mach Number in compressible flow theory. If Fr = 1, the flow is Critical.

The derivation has used a general uniform channel cross section in order to relate the definition to the general cross sections found in building drainage systems; however as the fundamental concepts remain identical in all cross section channels or partially filled pipeflows, it may be shown that for a uniform channel of rectangular cross section the wave speed above may be defined as

$$c = \sqrt{gh} \qquad (2.4)$$

Thus the wave speed is dependent upon the flow depth and the cross-sectional shape of the channel. Figure 2.3 illustrates the practical effect of this and implies that a wave propagating along a uniform channel will change shape during its propagation due to the variation in wave speed across the wave. Figure 2.3 also presents values of wave speed for a circular–cross section partially filled pipe flow; note that the wave speed is independent of pipe slope and roughness.

The variation of wave speed with flow depth is one of the most important characteristics of unsteady flow in building drainage systems, as it introduces the concept of wave attenuation. Waves are commonly observed to 'die away' along the length of a drain – this effect is entirely due to the variation in wave speed with depth, as the effect is to reduce peak flows and steepen leading edges of a wave while the trailing edge of the wave spreads. Figure 2.4 illustrates typical wave attenuation and identifies the main drivers. Wave attenuation is pronounced for appliance discharges featuring a rapidly changing profile – such as w.c. discharges – and is at a minimum for discharges featuring a 'plateau' profile – such as baths or showers. Wave attenuation is a major determinant of waste solid transport in a drain as the wave eventually attenuates to the state where it has insufficient energy or depth to maintain solid transport. Similarly it determines the maximum possible transport following a series of identical w.c. operations and the position in the channel beyond which transport becomes impossible without

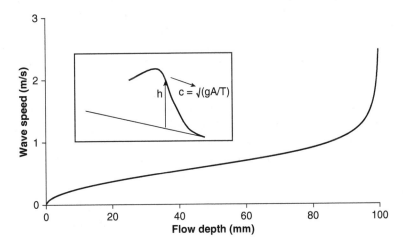

Figure 2.3 Dependence of wave speed on flow depth, illustrated for a partially filled circular–cross section channel (100 mm diameter).

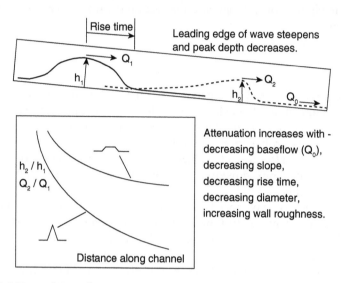

Figure 2.4 Dependence of attenuation on wave, flow and channel properties.

other joining flows. This will be returned to later in a discussion of solid transport simulation.

The establishment of steady uniform flow in building drainage system partially filled pipes and open channels requires that the potential energy transfer inherent in the channel downwards slope equates to the frictional 'losses'. Both laminar and turbulent flow conditions may occur in free surface channel flows; however laminar flow is rare due to the prevailing Reynolds Number of the flow. In full bore pipe flow it is accepted that the transition from laminar to turbulent flows occurs at Reynolds Number values above 2000, defined in terms of the fluid density and viscosity, the pipe diameter and the flow mean velocity as $Re = \rho u D/\mu$. The corresponding limit in free surface flow is defined by the Reynolds Number expression $Re = \rho u m/\mu$, where m is the hydraulic mean depth A/P, and the limit of laminar flow is set at around 500. In laminar flow, as the name suggests layers of fluid are assumed to slide over each other with the layer velocity rising away from the surface flowed over. In turbulent flow fluid particles progress in a random manner with an imposed mean velocity in the flow direction. Thus in turbulent flows higher velocity fluid is drawn into the fluid layers closer to the conduit wall, and therefore this type of flow displays 'flatter' velocity profiles across the flow than under laminar conditions.

The Reynolds Number limiting value may also be applied to non-circular free surface flow conditions if the non-dimensional term definition is recast as

Re = density × Characteristic velocity × Characteristic length / Dynamic Viscosity with the flow mean velocity remaining the characteristic velocity but the characteristic length defining the flow cross section being taken as the hydraulic mean depth, also known as the Hydraulic Radius, rather than the pipe diameter, where the hydraulic mean depth is defined as m = A/P, where A is the flow cross section and P is the wetted perimeter of the channel – note this will vary with flow depth in free surface flow applications. For full bore pipe flow the value of m becomes (πD/4)/πD or D/4 so that the limiting value of Reynolds Number in this form becomes 500. For open channels this value is only achieved for very shallow flows or where the μ/ρ ratio is very high. Generally therefore free surface flows are regarded as turbulent, a result that has consequences for the representation of frictional losses.

The fundamental equations of continuity, momentum and energy apply to free surface flows in uniform but general cross section conduits (Figure 2.5).

The continuity equation may be expressed between sections 1 and 2, Figure 2.5 as

$$Q = u_1 A_1 = u_2 A_2 \tag{2.5}$$

as the flow density may be taken as constant.

The momentum equation, or Newton's Second Law, may be expressed between sections 1 and 2, Figure 2.5, by equating the difference in hydrostatic forces to the rate of change of momentum between the two sections and including the gravitational force in the flow direction and the surface shear force resisting motion. The hydrostatic force driving the flow acceleration may be expressed as $(\rho g A \bar{y})_1 - (\rho g A \bar{y})_2$, where $\bar{y}$ is the depth of the flow centroid below the free surface.

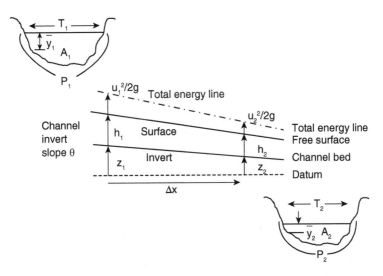

Figure 2.5 Steady non-uniform free surface flow in a uniform conduit.

The gravitational force in the flow direction may be expressed as the weight component in the flow direction, $mg\Delta x \sin\theta$.

The frictional resistance to flow is represented by the conduit surface shear stress acting over the wetted conduit wall area, $\tau_0 P\Delta x$, where τ_0 is the wall shear stress.

Thus the momentum equation becomes

$$(\rho g A\bar{y})_1 - (\rho g A\bar{y})_2 + mg\sin\theta - \tau_0 P\Delta x = \rho Q(u_2 - u_1) \tag{2.6}$$

It is usual to approximate the shear stress in terms of a surface friction factor, f, so that

$$\tau_0 = f\frac{1}{2}\rho u^2 \tag{2.7}$$

However it should be noted that there is a divergence between UK and US practice in defining the value of friction factor – this text will use the UK version.

The term $\rho g A\bar{y} + \rho Q u$ is referred to as the Specific Force, SF.

The Steady Flow Energy Equation defining steady non-uniform free surface flow may be determined by equating the potential and kinetic energy at each section and including an internal energy, or surrogate loss, term to represent frictional resistance; thus, for a flow with constant density

$$z_1 + h_1 + u_1^2 / 2g = z_2 + h_2 + u_2^2 / 2g + h_f \tag{2.8}$$

where h_f is the frictional loss term expressed in units of head of flowing fluid.

In the special case of steady uniform flow the velocity and depth remain constant in the flow direction so that

$$h_f = z_1 - z_2 \tag{2.9}$$

or the frictional resistance to the flow is counterbalanced by the potential energy change provided by the channel bed slope.

The combination of depth potential energy and flow kinetic energy terms, $h + u^2 / 2g$, is referred to as the flow Specific Energy, S.

The derivation of the continuity, momentum and energy equations assumes a mean flow velocity within the channel. In practice this is not the case, as there is a velocity distribution within the flow that features a particular maximum flow velocity depth dependent primarily on the channel cross-sectional shape. At the liquid-to-wall boundaries the condition of 'no-slip' imposes a zero local flow velocity; the accompanying shear forces generate the frictional resistance to flow to be represented by the Chezy[1] and Manning[2] expressions.

At the liquid-to-air interface the condition of 'no-slip' implies that the air immediately above the free fluid surface moves at the fluid surface velocity, thereby providing a shear force that entrains an airflow. In horizontal channels this is not a major issue, as the shear forces acting between the air and the 'dry' channel walls above the water free surface contribute a resistive shear force, and air entrainment is not significant. In annular stack flow the central air core is surrounded by an annular water flow so that the shear forces generated by the no-slip condition result in the air entrainment and subsequent air pressure regime within building drainage and vent systems (Swaffield 2010). Figure 2.6 illustrates a velocity profile across a flow within a partially filled pipe. The occurrence of maximum velocity at a depth below the surface has an impact on the transport of floating solids that will be returned to in the discussion of solid transport.

2.2 Frictional representation in steady uniform free surface flows

In the case of steady uniform flow the momentum equation, 2.6, may be simplified to $mg\Delta x \sin\theta = \tau_0 P\Delta x$ and further by introducing the friction factor relationship, equation 2.7, and replacing the mass term and the channel slope, so that

$$(\rho A\Delta x)g\Delta x \frac{\Delta z}{\Delta x} = f\rho\frac{1}{2}u^2 P\Delta x$$

$$\Delta z = f\frac{1}{2g}u^2\frac{P}{A}\Delta x = f\frac{1}{2gm}u^2\Delta x$$

where m is the hydraulic mean depth, A/P. For steady uniform flow it has been shown that the drop in channel invert equates to the frictional loss, h_f, so that

$$h_f = f\frac{1}{2gm}u^2\Delta x \qquad\qquad (2.10)$$

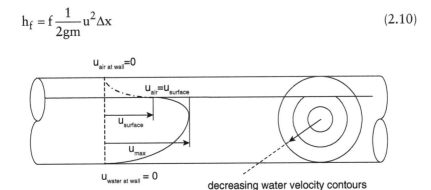

Figure 2.6 Water and air velocity profiles in a partially filled pipe flow.

which is the Darcy Equation with the diameter term replaced by the hydraulic mean depth and with the frictional loss being expressed as a head loss in m, historically a more appropriate measure in view of the application of the analysis to canal construction.

The channel bed slope S_0 may also be expressed as $h_f / \Delta x$ so that

$$Q = uA = A\sqrt{\frac{2g}{f}}\sqrt{mS_0} = AC\sqrt{mS_0} \qquad (2.11)$$

Equation 2.11 is the Chezy equation and is the basic expression defining steady uniform flow in free surface flows in channels and partially filled pipe flows. The coefficient C is the Chezy loss coefficient equivalent to the $\sqrt{(2g/f)}$ term and therefore has dimensions of $L^{0.5}T^{-1}$ so that the Chezy coefficient has a value and units dependent upon the system used, in the UK metric system $m^{0.5}\ s^{-1}$, in the US system $ft^{0.5}\ s^{-1}$.

Since its introduction in 1776 there have been numerous efforts to define the value of C for a range of channel conditions. A number of historical empirical relationships were proposed to define the Chezy coefficient, and while these should now be consigned to history, they still surface in some drainage design codes. In 1869 Ganguillet and Kutter proposed that, in SI units,

$$C = 23 + 0.00155 / S_0 + 1 / n) / (1 + (23 + 0.00155 / S_0)n / \sqrt{m} \qquad (2.12)$$

There has been some doubt as to the accuracy of the bed slope S_0 terms in equation 2.12.

The simplest and most well proven representation of C is due to Robert Manning, who in 1890 showed experimentally that C varied with $m^{1/6}$ and was dependent upon the roughness coefficient n of the channel walls. The Manning formula, developed by replacing C by his formulation, is usually written as

$$Q = ACm^{3/6}S_0^{0.5} = A\frac{m^{1/6}}{n}m^{3/6}S_0^{0.5} = A\frac{m^{0.667}}{n}S_0^{0.5} \qquad (2.13)$$

This expression, generally known as Manning's equation, is also sometimes referred to as the Strickler formula, and the term 1/n as the Strickler coefficient.

The coefficient n in equation 2.12, normally referred to as the Kutter equation, is identical to the Manning n developed 20 years later.

The Bazin formula published in 1897 does not relate C to the bed slope, S_0, and is expressed in SI units as

$$C = 86.9 / (1 + k / \sqrt{m}) \qquad (2.14)$$

where k is dependent upon channel roughness.

In common with full bore pipe flow, the value of C will vary with the flow Reynolds Number and the wall surface roughness. In this case the Reynolds Number is calculated based on mean flow velocity and hydraulic mean depth, m, as mentioned earlier. There is some experimental evidence that C does vary with cross-sectional shape and also with channel slope so that the mean flow velocity takes the form $u = Km^x S_0^y$, where K, x, and y are constants. The ASME 1963 report on channel frictional resistance concluded that the behaviour of the Chezy C term could be inferred directly from the full bore friction factor where the appropriate value of friction factor was determined from the Colebrook-White expression when the full bore pipe diameter was replaced with the free surface flow hydraulic mean depth, as shown by equation 2.15:

$$\frac{1}{\sqrt{f}} = -4\log_{10}\left(\frac{k}{14.8m} + \frac{0.315}{Re\sqrt{f}}\right) \qquad (2.15)$$

where Re is defined as $\rho um/\mu$ and k is the conduit wall surface roughness.

Tables 2.1 to 2.3 detail typical values of the Colebrook-White roughness coefficient, Manning's n and the Bazin coefficient, while Figures 2.7 and 2.8 illustrate the dependence of the Chezy coefficient on Reynolds Number and channel relative roughness and on Manning n for a 50% full bore partially filled pipeflow.

Ideally the ability to determine flow depth in a partially filled pipe flow for any given pipe slope and wall roughness would be desirable. Combining the Colebrook-White equation with the Chezy/Manning expression, as well as being able to define the flow Reynolds Number in terms of hydraulic mean depth, offers this possibility.

From the Manning equation $Q = A\sqrt{\frac{2g}{f}}\sqrt{mS_0}$, so that $\frac{1}{\sqrt{f}} = \frac{Q}{A\sqrt{2gmS_0}}$.

From the definition of Reynolds Number $Re = \frac{\rho um}{\mu} = \frac{\rho}{\mu}\frac{Q}{A}m$.

Table 2.1 Values of surface roughness, k mm, appropriate to the Colebrook-White expression, for a range of drainage pipe materials

Conduit surface material	Surface roughness k mm
Cast iron (coated)	0.15
Cast iron (uncoated)	0.30
Concrete	0.15
Glazed paper	0.06
UPVC	0.06
Glass	0.03

Table 2.2 Values of Manning n appropriate to the Chezy equation for a range of conduit, channel and pipe surface roughness

Conduit surface material	Good condition	Poor condition
Neat cement	0.010	0.013
Cement mortar	0.011	0.015
Concrete *in situ*	0.012	0.018
Concrete precast	0.011	0.013
Cement rubble	0.017	0.030
Dry rubble	0.025	0.035
Brick with cement mortar	0.012	0.017
Plank flume, planed	0.010	0.014
Plank flume, unplaned	0.011	0.015
Metal flume, semicircular, smooth	0.011	0.015
Metal flume, semicircular, corrugated	0.022	0.030
Cast iron, coated	0.012	0.015
Cast iron, uncoated	0.018	0.020
Steel riveted	0.017	0.025
Clay	0.014	0.018
UPVC	0.010	0.011
Glass	0.009	0.011
Canals, dredged, rock cut smooth or jagged, or rough beds with weeds	0.017 to 0.025 or 0.035	0.025 to 0.033 or 0.045
Natural streams, clean and smooth, rough or very weedy	0.025 to 0.045 or 0.075	0.035 to 0.060 or 0.150

Source: Douglas *et al.* 2005.

Table 2.3 Values of surface roughness, k, appropriate to the Bazin expression, for a range of general channel materials

Channel surface	Surface roughness k (mm)
Smooth cement or planed wood	0.060
Planks, ashlar and brick	0.160
Rubble masonry	0.460
Earth channels with rough surfaces	0.850
Earth channels	1.303
Exceptionally rough channels	1.750

Source: Douglas *et al.* 2005.

These values of f and Re may now be substituted into the Colebrook-White equation for free surface flow:

$$\frac{1}{\sqrt{f}} = -4\log_{10}\left(\frac{k}{14.8m} + \frac{0.315}{Re\sqrt{f}}\right)$$

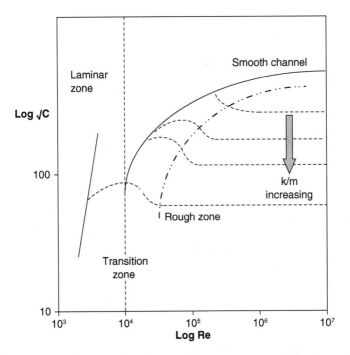

Figure 2.7 The dependence of the Chezy coefficient on Reynolds Number and channel relative roughness.

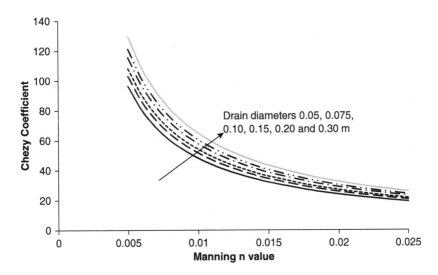

Figure 2.8 Values of the Chezy Coefficient based on Manning n for 50% full bore flow partially filled pipe flow for a range of typical building drainage diameters.

$$\frac{Q}{A\sqrt{2gmS_0}} = -4\log_{10}\left(\frac{k}{14.8m} + \frac{0.315}{\frac{\rho}{\mu}\frac{Q}{A}m\frac{A\sqrt{2gmS_0}}{Q}}\right)$$

$$Q = -4A\sqrt{2gmS_0}\log_{10}\left(\frac{k}{14.8m} + \frac{0.315}{\frac{\rho}{\mu}m\sqrt{2gmS_0}}\right)\qquad(2.16)$$

Equation 2.16 allows the flowrate to be determined at any depth, at any slope in a partially filled pipe flow with any pipe wall roughness. Figure 2.9 illustrates its application, while Figure 2.10 compares values of 50% full bore flowrate determined by either the Colebrook-White expression or Chezy utilising the Manning n coefficient.

Table 2.4 illustrates the 50% full bore flow in a range of pipe diameters and for a range of pipe slopes. The format of Table 2.4 is frequently met in national codes as it allows the flowrate/diameter/slope to be chosen so that a particular flow may be carried at no greater depth than 50% of the drain diameter. In most national codes the explicit flowrate presented in Table 2.4 is replaced by a Fixture or Discharge Unit value, a terminology that allows for the frequency of appliance usage to be included in the design.

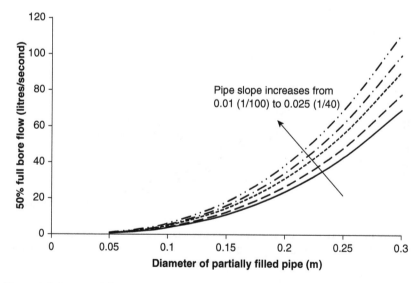

Figure 2.9 Increased flow capacity, 50% full bore, as the pipe slope is increased, flow predictions based on Colebrook-White with a wall roughness k of 0.06 mm.

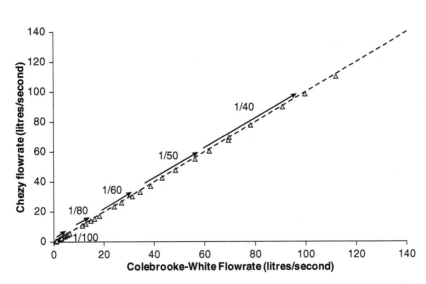

Figure 2.10 50% full bore flow capacity comparison between Colebrook-White predictions, with a wall roughness k = 0.06 mm, and Chezy predictions with a Manning's n of 0.009.

Table 2.4 Use of the Colebrook-White 50% full bore flow capacity variation with pipe slope and diameter as a basis for design tables indicating the maximum rating for any slope diameter combination

Pipe Diameter, m.	Pipe slope				
	0.01 1/100	0.0125 1/80	0.01667 1/60	0.02 1/50	0.025 1/40
0.05	0.60	0.68	0.79	0.87	0.98
0.075	1.77	2.00	2.33	2.57	2.89
	To increase allowable flow, increase pipe slope				
	To increase allowable flow, increase pipe diameter				
0.1	3.81	4.29	5.00	5.50	6.19
0.15	11.17	12.57	14.62	16.08	18.07
0.2	23.89	26.86	31.21	34.32	38.53
0.25	43.02	48.33	56.14	61.70	69.24
0.3	69.49	78.04	90.60	99.55	111.70

Note: Tabular data of this form common in historic drainage design guides.

2.3 Optimum flow depth in partially filled pipeflow

Partially filled pipeflow rate and mean velocity at any flow depth depends, as shown by the Chezy equation, on the value of the hydraulic mean depth at that degree of pipe full bore flow. Due to the geometry of a circular–cross section drain there will therefore be an optimum depth at which the flowrate is a maximum and a second optimum flow depth at which the mean flow velocity will be a maximum. Figures 2.11 and 2.12 illustrate the geometry for a circular–cross section drain and the variation of flow area, wetted perimeter, surface width and hydraulic mean depth with flow depth.

As the depth approaches full bore, the wetted perimeter rises more rapidly than the flow cross-section area, so that the hydraulic mean depth falls and the flowrate and mean velocity decrease. Figure 2.13 illustrates the resultant optimum depth for maximum mean velocity, 0.81 D, and maximum flowrate, 0.95 D.

These optimum values may be confirmed by reference to the flow equations:

$$u = Cm^{1/2}S_0^{1/2} = C\left(\frac{A}{P}\right)^{1/2} S_0^{1/2}$$

So that for a constant slope and Chezy Coefficient, it follows that the maximum value of u will occur when A/P is a maximum; hence for maximum velocity

$$\frac{d(A/P)}{d\theta} = \left(\frac{1}{P^2}\left(P\frac{dA}{d\theta}\right) - A\frac{dP}{d\theta}\right) = 0$$

$$P\frac{dA}{d\theta} = A\frac{dP}{d\theta} \tag{2.17}$$

Substituting from Figure 2.11 yields

$$2r\theta(1 - \cos 2\theta)r^2 = r^2(\theta - 0.5\sin 2\theta)2r$$
$$\theta(1 - 2\cos\theta) = \theta - 0.5\sin 2\theta$$

$2\theta = \tan 2\theta$, so that $2\theta = 257.5°$, and the flow depth becomes $r(1 - \cos\theta)$ or 0.81 D.

For constant slope and Chezy Coefficient, the flowrate Q will be a maximum when (A^3/P) is a maximum, so that

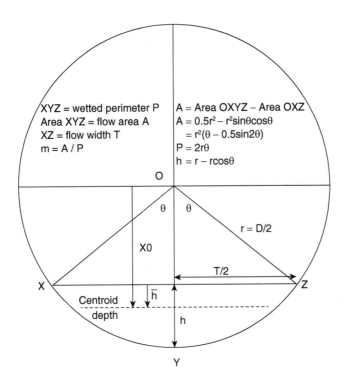

Figure 2.11 Geometry of a circular–cross section drain.

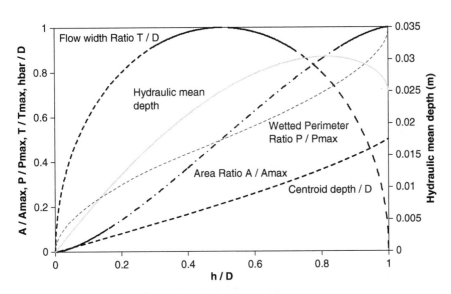

Figure 2.12 Geometrical variation of area, wetted perimeter, flow width and hydraulic mean depth for a circular cross section partially filled drain flow.

$$\frac{d(A^3/P)}{d\theta} = \left(\frac{1}{P^2}\left(3PA^2\frac{dA}{d\theta}\right) - A^3\frac{dP}{d\theta}\right) = 0$$

$$3P\frac{dA}{d\theta} = A\frac{dP}{d\theta} \tag{2.18}$$

Substituting from Figure 2.11 yields

$$6r\theta(1-\cos 2\theta)r^2 = r^2(\theta - 0.5\sin 2\theta)2r$$

$$4\theta - 6\theta\cos 2\theta + \sin 2\theta \tag{2.19}$$

so that $2\theta = 308°$, and the flow depth becomes $r(1 - \cos\theta)$ or 0.95 D.

Figures 2.8 to 2.9 may be replicated for any channel or partially filled pipe numerically, as will be demonstrated in a later treatment of non-circular section drains running partially filled, Cummings *et al.* 2007. Based on the following equations, Figure 2.13 illustrates the variation in flow mean velocity and flowrate with depth for a circular–cross section drain, confirming the analysis developed above.

The full bore flowrate is some 94% of the maximum flowrate achieved with a partially filled pipe.

This numerical approach will be applied later to examples including parabolic cross sections and experimental sections considered by Cummings in the Caroma study as the θ, A, P and T expressions may be replaced by equations appropriate to the new cross-sectional shape:

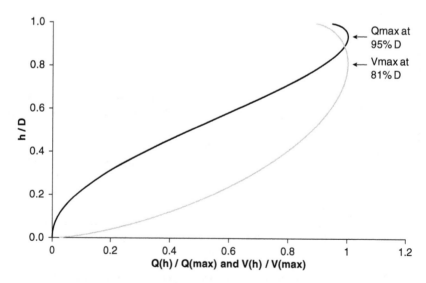

Figure 2.13 Variation of flow mean velocity and flowrate with depth for a circular cross section.

h < r	then	$\theta = 2\,\mathrm{atan}(\sqrt{(h - (D - h))}/(D/2 - h))$
h = r	then	$\theta = \pi$
h > r	then	$\theta = \pi + 2\,\mathrm{atan}((h - r)/(\sqrt{(h(D - h))}))$

$$A = ((D^2/8)(\theta - \sin\theta)$$
$$P = D\theta/2$$
$$T = 2\sqrt{(h(D - h))} \qquad\qquad (2.20)$$
$$m = A/P$$
$$Q = A(m^{0.667})\sqrt{S_0}/n$$
$$V = Q/A$$

2.4 Normal and Critical depths in free surface channel flow

Inherent in the steady uniform channel flow treatment is the concept of Normal depth flow – characterised by the Chezy Equation, whether the Chezy Coefficient is defined in terms of Manning n or the Colebrook-White friction relationship.

For a uniform section channel at a constant slope and with constant wall roughness, flow depth at any applied flowrate will be given by the Chezy Equation in the form

$$f(h) = 1 - \frac{Q}{CA_n\sqrt{m_n S_0}} = 0 \qquad\qquad (2.21)$$

The function will be zero when the flow depth h corresponds to the required Normal flow depth.

The Critical depth, defining the transition from subcritical to supercritical flow or vice versa may be defined by a function of the form

$$f(h) = 1 - \frac{Q^2 T_c}{gA_c^3} = 0 \qquad\qquad (2.22)$$

The function will be zero when the flow depth corresponds to the Critical depth.

Values of Normal and Critical depth, easily determined via the bisection technique,[3] are essential in establishing the pretransient steady flow conditions in any building drainage network and, in the case of Critical depth, defining system boundary conditions.

Figure 2.14 illustrates the variation of Normal depth with flowrate in a 100 mm diameter partially filled pipe for a range of channel slopes, note in particular the maximum flow rate at less than full bore flow, as suggested by Figure 2.13. The Critical depth is independent of channel slope, and one Critical depth curve defines the regime boundary between subcritical to supercritical flow. Figure 2.15 illustrates the variation of Normal and Critical depth with drain diameter for a single channel slope of 0.01.

Figure 2.16 shows the effect of increasing roughness. As the Manning's n value increases so the Normal flow condition moves from supercritical to subcritical. This transition, which may occur with age as drains become rougher due to deposits and corrosion may be the explanation for the belief that hydraulic jumps occur at entry to horizontal drains from a building vertical stack, an effect not readily replicated in the laboratory.

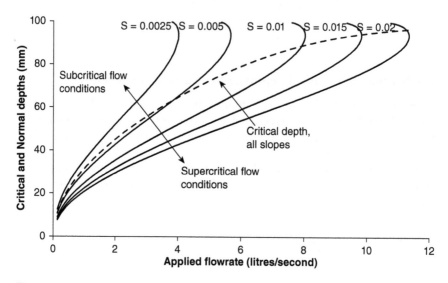

Figure 2.14 Variation of Normal depth with channel slope and applied flow for a 100 mm diameter drain.

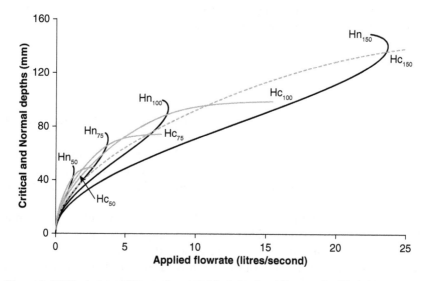

Figure 2.15 Variation of Normal and Critical depth with partially filled drain diameter and applied flow at a slope of 0.01.

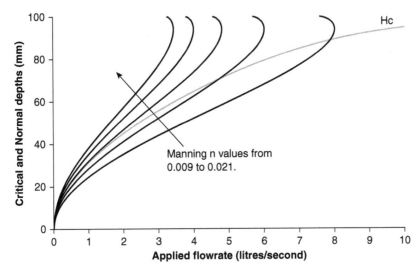

Figure 2.16 Variation of Normal and Critical depth with Manning n value for a 100 mm diameter drain at 0.01 slope.

2.5 Steady non-uniform free surface flows – gradually varied flow profiles

The equations developed for frictional resistance and channel flowrate and mean velocity are applicable to the steady uniform flow experienced in long uniform channels of constant slope subject to a constant inflow. However channel slope and cross section may change in the flow direction, so that while the flowrate may remain constant with time – steady flow – the local parameters of mean flow velocity and depth may well change in the flow direction. Such flow conditions are referred to as steady non-uniform flow.

The flow Specific Energy was earlier defined as the sum of the flow kinetic and potential energy values, namely

$$SE = h + u^2 / 2g \qquad (2.23)$$

Equation 2.23 may be expressed in terms of the flowrate so that

$$SE = h + (Q / A_h)^2 / 2g \qquad (2.24)$$

where the flow area A is a function of the flow depth h.

It is apparent from equation 2.24 that there are two solutions in h that satisfy the equation, illustrated in Figure 2.17. Note that subcritical flow is characterised by deep flow at relatively low mean velocities, while the supercritical flow has higher mean velocity values and shallower depth. The boundary is formed by the Critical depth which is independent of pipe slope and roughness. The Critical depth may be determined from Figure 2.17 and equation 2.22 as the Critical depth corresponds to both the minimum Specific Energy and the maximum flowrate.

Hence from equation 2.24, at the Critical depth, it follows that

$$\frac{dSE}{dh} = \left(1 - \frac{2Q^2}{2gA^2}\frac{dA}{dh}\right)_c = 0 \tag{2.25}$$

A small increase in flow area dA may be equated to Tdh, and so equation 2.25 yields

$$1 - \frac{Q^2 T_c}{gA_c^3} = 0$$

If Critical conditions are designated by suffix c, then the Critical velocity defining the transition from subcritical to supercritical flow may be seen to be

$$u_c = \frac{Q_c}{A_c} = \sqrt{\frac{gA_c}{T_c}} \tag{2.26}$$

It will be seen that this velocity has already been identified in general as the velocity of propagation of a surface wave dependent upon the

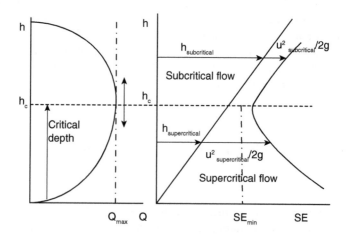

Figure 2.17 Relationships between flowrate Q, Specific Energy, SE, and the boundaries of subcritical and supercritical flow defined in terms of the flow Critical depth.

flow depth. Hence at the Critical velocity no wave may be transmitted upstream as the flow velocity equates to the wave propagation velocity. This is a further definition of the boundary between subcritical and supercritical flows.

It should also be noted that the Critical velocity and Critical depth are independent of channel slope. This becomes an important boundary condition in the later application of the Method of Characteristics to solve the unsteady flow regimes met in building drainage networks.

In general the predominant flow condition in building drainage network free surface flow is supercritical flow. This observation is however tempered by the fact that the natural obstructions to flow, due to either defective drainage installation, deposits or the effect of joining flows at junctions, impose local depths that may exceed the locally appropriate Critical depth. This forces a zone of subcritical flow onto the system that extends upstream until terminated by a hydraulic jump – effectively a shock that transforms the approaching supercritical flow into the deeper subcritical flow that comprises a backwater profile immediately upstream of the obstruction, drain defect or flow confluence at a junction. Conversely, downstream of any obstruction or junction generating a change in approach flow regime from supercritical to subcritical there will be a zone where the flow conditions revert to supercritical, the flow depth downstream of the obstruction or junction passing through the Critical depth value. This provides a useful entry condition for the drain sections downstream of a junction.

It is therefore necessary to determine the flow depth in the transition zones upstream or downstream of a local obstruction or junction. The flow in these areas is referred to as steady non-uniform or gradually varied flow. The transition across a hydraulic jump is referred to as rapidly varying flow and will be dealt with independently later.

Figure 2.3 has already defined the steady non-uniform flow regime. Gradually varied flow is defined as identical to this regime under the condition that the flow conditions change relatively slowly in the flow direction so that at any location it is acceptable to represent the frictional loss over any small distance increment Δx by its equivalent steady state values at the local values of flow depth and velocity and calculated through the Chezy equation. Thus, relative to some datum channel invert level Z_0 set $S_0 \Delta x$ upstream, the energy slope of the channel under steady non-uniform conditions may be seen from Figure 2.3 and equation 2.4 to be

$$S = \frac{d}{dx}\left(\frac{u_1^2}{2g} + (Z_0 - S_0 \Delta x) + h \right) \qquad (2.27)$$

Note that under steady uniform flow conditions the invert slope S_0 is identical to the friction slope S so that applying the Chezy equation to local values of the channel and flow parameters yields the local value of S, namely

$$S = Q^2 / C^2 mA^2 \tag{2.28}$$

Thus from equations 2.27 and 2.28

$$S = \frac{d}{dx}\left(\frac{u_1^2}{2g} + (Z_0 - S_0 \Delta x) + h\right) = \frac{u}{g}\frac{du}{dx} - S_0 + \frac{dh}{dx} = -\frac{Q^2}{C^2 mA^2} \tag{2.29}$$

As the flow is steady dQ/dx = 0 and as Q = uA and dA = Tdh as before, it follows that

$$\frac{dQ}{dx} = \frac{duA}{dx} = A\frac{du}{dx} + u\frac{dA}{dx} = 0$$

and

$$\frac{du}{dx} = -\frac{u}{A}\frac{dA}{dx} = -\frac{uT}{A}\frac{dh}{dx} = -\frac{QT}{A^2}\frac{dh}{dx}$$

Substituting into equation 2.29 yields

$$\frac{u}{g}\frac{QT}{A^2}\frac{dh}{dx} + S_0 - \frac{dh}{dx} = \frac{Q^2}{C^2 mA^2}$$

$$\frac{dh}{dx}\left(\frac{Q^2 T}{gA^3} - 1\right) = \frac{Q^2}{C^2 mA^2} - S_0$$

$$dx = \frac{\left(\frac{Q^2 T}{gA^3} - 1\right)}{\frac{Q^2}{C^2 mA^2} - S_0} dh = \frac{1 - Q^2 T / gA^3}{S_0 - Q^2 / C^2 mA^2} dh \tag{2.30}$$

The distance along the channel to achieve a depth change dx is therefore given by the integral of equation 2.30:

$$dx_{h_1 - h_2} = \int_{h_1}^{h_2} \frac{1 - Q^2 T / gA^3}{S_0 - Q^2 / C^2 mA^2} dh = \int_{h_1}^{h_2} \frac{1 - Fr^2}{S_0 - S_f} dh \tag{2.31}$$

It will be appreciated that the numerator of equation 2.31 is the expression for Critical depth and the denominator the expression for Normal depth, both developed earlier. Further it will be seen that if the denominator tends to zero then the flow is steady uniform and there is no change in flow depth in the flow direction, namely Normal depth steady flow conditions corresponding to the channel slope, cross-sectional dimension and roughness. Similarly if the numerator tends to zero then the flow is Critical and there is no change in x for a change in h, the condition already defined as a hydraulic jump, equivalent to a shock wave.

The alternative form involves inclusion of the flow Froude Number, Fr, and the channel friction slope S_f, defined previously in equations 2.3 and 2.28

While for most building drainage systems the channel cross sections will not be rectangular, it will be appreciated that for rectangular sections the expression may be cast in terms of flow depth and channel width only. For regular but non-rectangular sections, either circular or some other desired cross section, equation 2.31 may be integrated to give the changing flow depth in the flow direction by means of Simpsons Rule.[4]

Examples of gradually varied flow profiles will be returned to later in this chapter as a precursor to defining the entry and exit conditions necessary to model unsteady flow within building drainage systems. In order to complete the necessary equations defining the free surface flow conditions in drainage networks it is also necessary to consider rapidly varying flow and in particular the properties of the hydraulic jump.

2.6 Rapidly varied flow – the hydraulic jump

The hydraulic jump is a crucially important example of local steady non-uniform flow. From equation 2.31 it would appear that the water surface would theoretically have to display a vertical transition from supercritical to subcritical flow conditions at the jump as Δx tends to zero; however this is not a supportable real situation, and in practice the depth change occurs over a short distance with a steeply sloping water surface accompanied by severe eddies, turbulence and a substantial loss of energy. Figure 2.18 illustrates the jump conditions in steady flow – i.e. the jump is stationary. As the mean velocity of the steady flow decreases downstream of the jump there will be an increase in depth and a hydrostatic force acting to oppose the flow and generate the jump. This set of conditions clearly excludes the assumption in the treatment of steady non-uniform gradually varied flow that the frictional losses experienced could be represented by the local flow parameters within the Chezy equation. The rate of change of the flow parameters is too great to support this simplification.

As the jump is stationary the continuity of flow equations is

$$u_1 A_1 = u_2 A_2 = Q$$

The equation of motion applied across the jump is similarly stated as

$$\rho g A_1 \bar{h}_1 - \rho g A_2 \bar{h}_2 + mg \sin \theta - \tau_0 P \Delta x = \rho Q (u_2 - u_1)$$

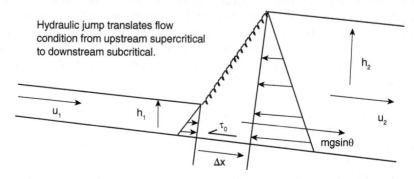

Figure 2.18 Forces acting across a hydraulic jump in steady partially filled pipe flow.

Generally it is assumed that Δx tends to zero, eliminating the shear force term, and that the channel slope may be ignored so that the equation of motion reduces to

$$\rho g A_1 \bar{h}_1 - \rho g A_2 \bar{h}_2 = \rho Q (u_2 - u_1) \qquad (2.32)$$

Rearranging equation 2.32 yields

$$SF_1 = \rho g A_1 \bar{h}_1 + \rho Q^2 / A_1 = SF_2 = \rho g A_2 \bar{h}_2 + \rho Q^2 / A_2$$

Equation 2.33 may be solved for the two sequent depths across the hydraulic jump, i.e. depths where the Specific Force values are identical from equation 2.33, by means of the bisection method introduced earlier.

The centroid depth for a partially filled pipe flow may be determined from the geometry of the circular cross section, Figure 2.11, as

$$\bar{h} = X0 + h - \frac{D}{2} = 0.666 \frac{D}{2} \frac{\left(3\sin\dfrac{\theta}{2} - \sin 3\dfrac{\theta}{2}\right)}{4\left(\dfrac{\theta}{2} - \dfrac{1}{2}\sin\theta\right)} + h - \frac{D}{2} \qquad (2.33)$$

where similar relationships may also be developed for non-circular cross sections. If the applied flow rate and the channel parameters of diameter, slope and roughness are known, then the channel Critical and Normal depths may be determined and the flow condition defined as subcritical or supercritical steady uniform flow.

Two important cases require to be addressed, dependent upon the control applied to the flow. Under subcritical flow conditions the flow control may be located downstream as surface waves may be propagated upstream. In supercritical flow the control is applied upstream.

Consider first a supercritical flow condition dependent upon channel slope and roughness where the channel is terminated by a junction or obstruction that generates a local flow depth in excess of the flow Critical depth. In this case the junction or obstruction generates a backwater surface profile that reduces flow depth upstream until a hydraulic jump is reached that allows the flow depth to decrease to the upstream normal supercritical value. Figure 2.19 illustrates this case, while Figure 2.20 presents sequent depth values for a 100 mm diameter drain set at a 0.01 slope with a Manning n surface roughness of 0.009.

The second case of particular importance is the formation of a hydraulic jump downstream of a channel entry where the entry flow depth is less than the Normal flow depth for that flowrate, channel cross-sectional dimensions and roughness (Figure 2.21). For a jump to form, the downstream flow must be subcritical and the entry depth supercritical. A water profile is then established downstream from the channel entry, and the flow depth increases until the sequent depth corresponding to a downstream normal subcritical flow is achieved. A jump forms that raises the water depth to the Normal flow depth in the particular drain being considered.

It will be appreciated that this can only occur if the downstream flow is subcritical. Many well-established texts show a hydraulic jump downstream of drain entry as an expected characteristic of drain flow; however it has been difficult to replicate this belief in the laboratory. Figure 2.22 shows that a jump will only be present if the drain is rough or at a very shallow gradient. As drain roughness increases with time it may well be that the believed presence of a drain entry jump is based on observations of flow in aged pipes or drains set at a very shallow gradient. Referring to Figure 2.16, it will be seen that the channel roughness would have to be at a Manning n value between 0.012 and 0.015 before the downstream subcritical conditions are present at the drain diameter and slope chosen. Figure 2.23 illustrates the dependence on gradient – the drain has a 0.009 Manning n value, but the slope has been decreased to 0.0025.

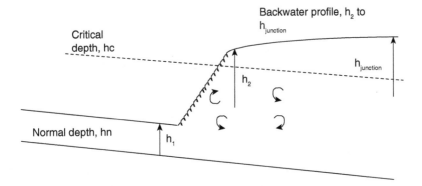

Figure 2.19 Hydraulic jump formed upstream of a junction or flow obstruction. Note sequent depths are the upstream Normal depth and the jump height.

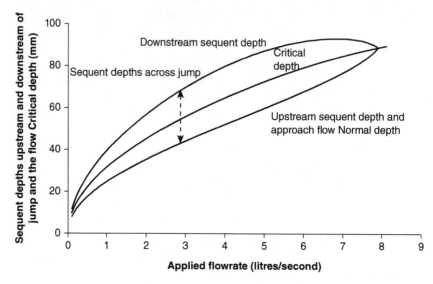

Figure 2.20 Sequent depth across a hydraulic jump formed upstream of a junction or flow obstruction in a 100 mm diameter drain at 0.01 slope with a Manning n value of 0.009.

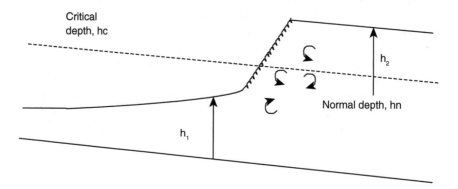

Figure 2.21 Hydraulic jump formed downstream of a channel entry. Note sequent depths are the downstream Normal depth and the flow depth immediately upstream of the jump.

2.7 Entry and exit flow depth boundary conditions for both subcritical and supercritical flows

While the predominant flow condition within building drainage systems will be supercritical, the presence of gradually varied flow depth profiles at drain entry and exit, and in the vicinity of hydraulic jumps, is an important element in establishing any steady flow regime prior to the imposition of unsteady flows due to random appliance discharges.

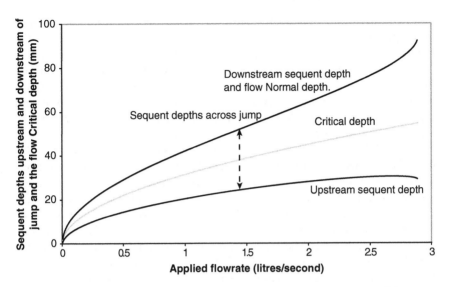

Figure 2.22 Sequent depth across a hydraulic jump formed downstream of drain entry in a 100 mm diameter drain at 0.01 slope with a Manning n value of 0.025.

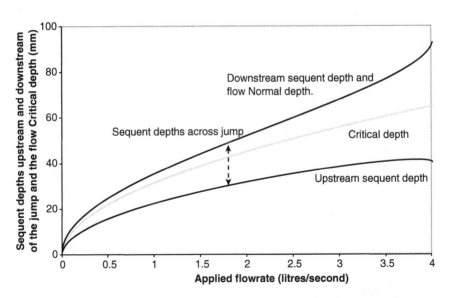

Figure 2.23 Sequent depth across a hydraulic jump formed downstream of drain entry in a 100 mm diameter drain at 0.0025 slope with a Manning n value of 0.009.

Two important boundaries must be addressed, namely flow depth development immediately downstream of a stack or junction entry to the downstream drainage network and the flow depth conditions immediately upstream of a free discharge to a vertical stack, to a junction of two or more branches or to an obstruction. In addition the presence of hydraulic jumps must be analysed, both in terms of their sequent or conjugate depths and their location within the flow – and therefore their influence on surface profiles upstream and downstream of their location.

Figure 2.24 illustrates the flow conditions at a stackbase where the annular stack flow translates into the free surface branch flow. The flow conditions at the base of the stack depend upon the kinetic energy of the annular flow, and the flow depth at drain entry is therefore likely to be less than the Critical depth for that flow in a drain with the diameter, roughness and slope of the branch. If the downstream flow is supercritical the boundary conditions for the gradually varied flow depth profile downstream from the stackbase may be seen to be the energy-defined stack exit depth and the downstream flow Normal depth. In this case, as shown by Figures 2.24 and 2.25, there is no hydraulic jump, and the gradually varied depth profile rises to become asymptotic to the flow Normal depth.

If the downstream flow is expected to be subcritical due to the branch slope or roughness, then it will be necessary for the stack discharge flow to pass through a hydraulic jump before the subcritical normal depth is achieved. Figure 2.26 illustrates this condition, while Figure 2.27 presents a Simpson's Rule integration of equation 2.31. It should be noted that if the downstream flow is controlled by a junction or defect imposed depth that exceeds the Critical value for the applied flowrate then the gradually varied flow profile downstream of the stack exit will rise to the imposed jump

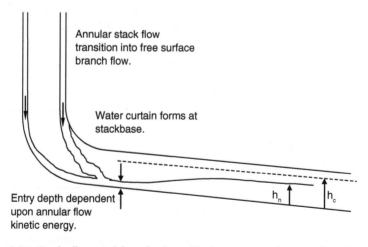

Figure 2.24 Gradually varied flow depth profile downstream of a vertical stack discharge to horizontal branch where the downstream flow is supercritical.

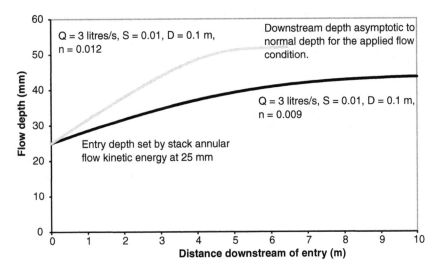

Figure 2.25 Simpson's Rule prediction of the gradually varied flow depth profile downstream of a vertical stack discharge to a horizontal branch under supercritical flow conditions.

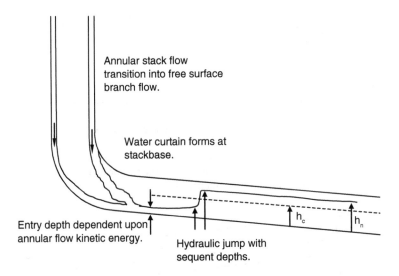

Figure 2.26 Gradually varied flow depth profile downstream of a vertical stack discharge to horizontal branch where the downstream flow is subcritical.

upstream conjugate depth. A second gradually varied flow profile will then be imposed downstream of the jump to the junction itself.

The conditions at branch exit are also determined by reference to gradually varied flow theory. In the case of supercritical flow in a drain featuring a free discharge to a free volume or to an empty vertical stack, it follows from the definition of supercritical flow that the flow remains 'unaware' of the exit condition. In this case the flow exits the drain at its Normal depth, as illustrated in Figure 2.28.

In the case of the discharge of subcritical flow via a free outfall the presence of the branch exit is communicated upstream as the flow accelerates over the lip of the drain termination. In this case a drawdown profile is established upstream of the discharge point. The flow passes through its Critical depth value approximately one drain diameter upstream of the discharge as shown in Figures 2.29 and 2.30.

Conditions at a junction of two or more branch horizontal drains is more complex. In this case the flow depth at the junction itself is determined by

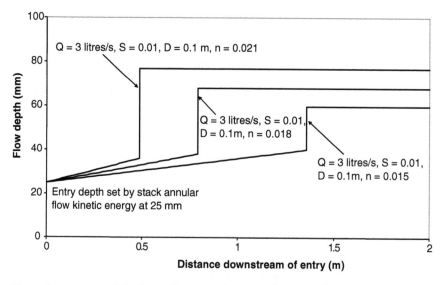

Figure 2.27 Simpson's Rule prediction of the gradually varied flow depth profile downstream of a vertical stack discharge to horizontal branch where the downstream flow is subcritical.

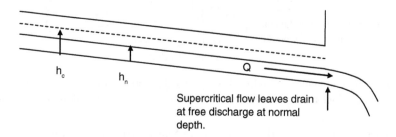

Figure 2.28 Supercritical flow exits a free discharge at its Normal depth as no indication of the presence of the exit can be transmitted upstream as the wave speed is less than the flow mean velocity.

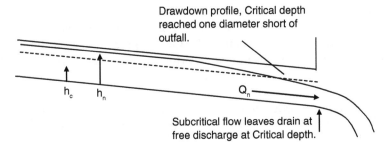

Figure 2.29 Subcritical flow exits a free discharge at its critical depth as information concerning the presence of the exit is transmitted upstream as the wave speed exceeds the flow mean velocity.

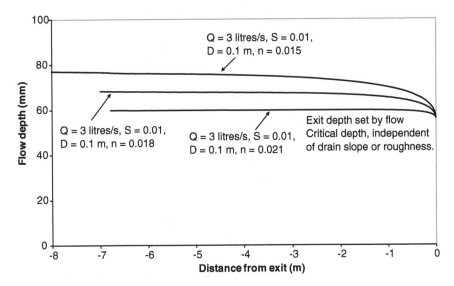

Figure 2.30 Simpson's Rule predictions of the subcritical flow depth profile upstream of a free outfall at Critical depth.

the junction geometry and the total of the approach flows in all the joining branches. The combined flow exits the junction at the downstream combined flow Critical depth when the flows are expected to be supercritical.

Hydraulic jumps are established in each of the approach branches where the local flow is supercritical, as illustrated in Figure 2.31. It should be noted that even if one of the connecting branches was carrying a minimal flow a jump would be established, as backflow from the junction would pass into the low flow branch and establish a backwater profile. Figure 2.31 illustrates this condition when all the drains are expected to carry supercritical flows, while Figure 2.32 presents a Simpson's Rule integration of equation 2.31 for differing approach flows in each of the joining branches.

In the case of a top entry junction, Figure 2.33, a hydraulic jump is only formed in the lower horizontal branch. The upper branch discharges to the junction at the appropriate supercritical normal depth for the flow carried. In the unlikely case of a top entry subcritical approach flow, the upper flow would discharge to the junction at its Critical depth value. Figure 2.33 illustrates the first and more common case.

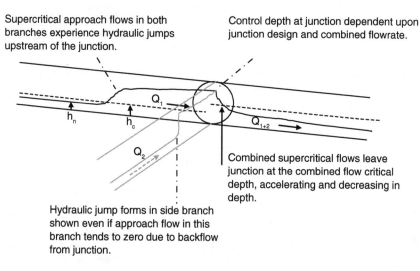

Figure 2.31 Hydraulic jumps established upstream of a multiple branch junction carrying supercritical flows.

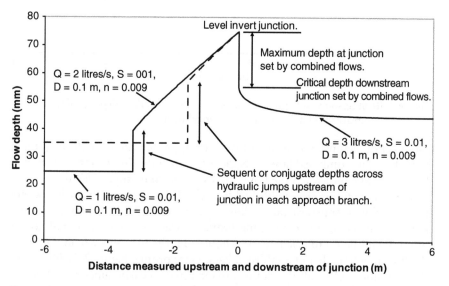

Figure 2.32 Simpson's Rule prediction of the depth profiles upstream and downstream of a level invert junction in a supercritical flow network.

If the approach flows to a level invert junction are subcritical, Figure 2.34, then backwater profiles form from the junction imposed depth to the approach flow Normal depths. Downstream flow again leaves the junction at the appropriate combined flow Critical depth.

It should be noted in reviewing the Simpson's Rule integrations presented that the water surface profiles appear steep graphically due to the disparity of the x and y axis dimensions, horizontally in m and vertically in mm. The backwater profile between the jump and the junction in Figure 2.32 has a

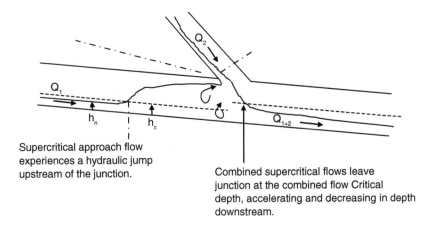

Supercritical approach flow experiences a hydraulic jump upstream of the junction.

Combined supercritical flows leave junction at the combined flow Critical depth, accelerating and decreasing in depth downstream.

Figure 2.33 Hydraulic jump established upstream of a top entry branch junction carrying supercritical flows.

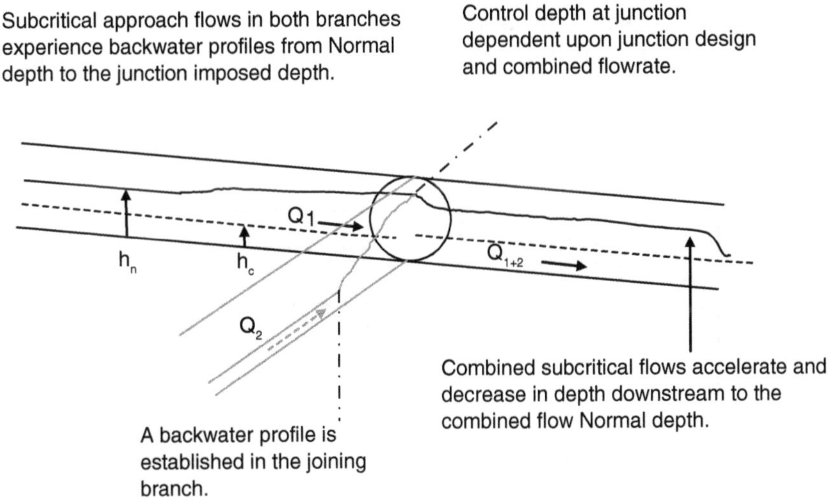

Subcritical approach flows in both branches experience backwater profiles from Normal depth to the junction imposed depth.

Control depth at junction dependent upon junction design and combined flowrate.

A backwater profile is established in the joining branch.

Combined subcritical flows accelerate and decrease in depth downstream to the combined flow Normal depth.

Figure 2.34 Backwater profiles established upstream of a multiple branch junction carrying subcritical flows.

slope of 1/100, so in practice, when the downward slope of the drain is taken into consideration, would appear almost horizontal in a laboratory or specialist drainage application glass pipe.

Thus by reference to the equations governing Normal and Critical depth and developing an integral expression, equation 2.31, to define the flow depth profiles governed by either upstream boundary conditions, in supercritical flow, or downstream boundary conditions, in subcritical flow, it has been possible to develop an understanding of the free surface flow regime that dominates within steady open channel/building drainage horizontal branch flows.

2.8 Annular free surface flows

The flow regime within the horizontal branches and sewer connections in a building drainage system will be predominantly supercritical free surface flow, with transitions, where imposed by local boundary conditions, to zones of subcritical flow. Rainwater gutter flow will be predominantly subcritical due to the gutter slopes commonly employed. However the flow in the vertical stacks connecting and collecting from individual floors will be annular, as will the rainwater flows in vertical downpipes – with the exception of siphonic rainwater system downpipes where the flow regime under design conditions will be full bore. It is therefore necessary to review the governing equations for annular free surface flows in order to provide the necessary interfloor linkages for any comprehensive system model.

Figure 2.35 illustrates a typical branch water discharge to a vertical stack. The flow tends to arch across the vertical stack to impinge on the far wall surface, imparting a swirl component to the flow velocity that allows the annular flow to be established. As the flow moves down the stack it accelerates and the annular film thins. Early researchers were surprised that the annular flow velocity became independent of stack height;

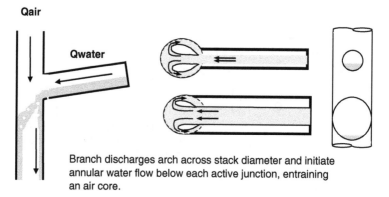

Branch discharges arch across stack diameter and initiate annular water flow below each active junction, entraining an air core.

Figure 2.35 Mechanism establishing annular flow below each active vertical stack to branch junction.

however this is easily explained in terms of a terminal velocity and a development length beyond which there is no further flow acceleration, illustrated by Figure 2.36

Under terminal conditions the flow acceleration is zero, so for an incremental length of stack, Δz, the gravity force, $\rho g \, \pi D t \, \Delta z$, must equal the wall shear stress, $\tau_0 \pi D \Delta z$. If the wall shear stress is expressed as $0.5 \rho f V_w^2$, where f is an appropriate friction factor, and it is assumed that the water to air shear may be neglected, then the terminal annular thickness, t_t and velocity, V_t, values become

$$t_t = (f / 2g)V_t^2$$

and

$$V_t = \sqrt{2gt_t / f)}$$

so that the friction factor f becomes

$$\frac{1}{\sqrt{f}} = \frac{V_t}{\sqrt{2gt}} = \frac{Q_w}{\pi Dt \sqrt{2gt}} \qquad (2.34)$$

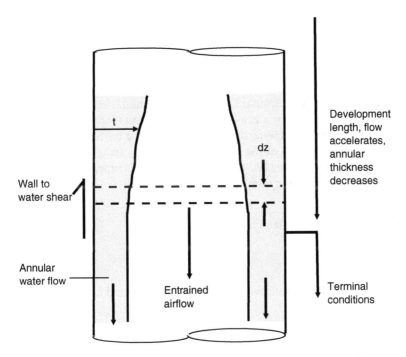

Figure 2.36 Forces acting on the annular film and the basis for a terminal flow condition once the annular water film reaches a terminal velocity.

The Colebrook-White friction factor expression may be applied to free surface flow so long as the characteristic length in both the Reynolds Number and roughness ratio is taken as the hydraulic mean depth, m = Area / Wetted Perimeter, equation 2.15

$$\frac{1}{\sqrt{f}} = -4\log_{10}\left[\left(\frac{k}{14.84m}\right) + \frac{0.313}{Re\sqrt{f}}\right]$$

where

$$m = \frac{\pi Dt}{\pi d} = t$$

and

$$Re = \frac{\rho V_t m}{\mu} = \frac{\rho V_t t}{\mu} = \frac{\rho Q_w}{\mu \pi D}$$

Substituting into the Colebrook-White equation above yields

$$\frac{Q_w}{\pi Dt\sqrt{2gt}} = -4\log_{10}\left[\left(\frac{k}{14.84t}\right) + \frac{0.313\mu}{\rho t\sqrt{2gt}}\right] \qquad (2.35)$$

For annular flows in stacks of known roughness equation 2.35 may be used to determine the terminal thickness and terminal velocity. Figure 2.37 represents such results for a smooth stack, k = 0, and identifies the maximum permitted flow, set by standards as that flow with a terminal annular thickness not exceeding 25% of the stack cross-sectional area, t < 1/16 D.

Under terminal conditions, $dV_t/dt' = 0$, where t' is time, it follows that the shear force balances the gravity force, hence

$$\rho g \pi Dt \Delta z = \frac{1}{2}\rho f V_t^2 \pi D \Delta z$$

and so as $t = \dfrac{Q_w}{\pi DV_t}$, it follows that $g\dfrac{Q_w}{\pi DV_t} = \dfrac{1}{2}fV_t^2$, and hence

$$V_t = \left(\frac{2g}{f}\frac{Q_w}{\pi D}\right)^{1/3}$$

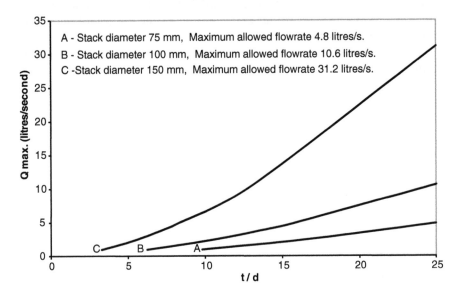

Figure 2.37 Flowrate in a smooth vertical stack from equation 2.35 with the maximum set when the annular flow area becomes 25% of the stack cross section.

From the Chezy equation it will be seen that $\sqrt{\dfrac{2g}{f}} = \dfrac{m^{1/6}}{n}$ where m is the hydraulic mean depth and n is the Manning Coefficient. For an annular flow taking up no more than 25% of the stack cross-section area, the hydraulic mean depth, m, will approximate the annular thickness, t and, as $t = Q_w / \pi D V_t$, the expression for terminal velocity becomes

$$V_t = \left[\frac{1}{n^2} \left(\frac{Q_w}{\pi D V_t} \right)^{1/3} \frac{Q_w}{\pi D} \right]^{1/3}$$

and

$$V_t^{10/9} = \left(\frac{1}{n^2} \frac{1}{\pi^{4/3}} \right)^{1/3} \left(\frac{Q_w}{D} \right)^{4/9}$$

so that, if $K = 0.632/n^{0.6}$, terminal velocity becomes

$$V_t = K \left(\frac{Q_w}{D} \right)^{0.4} \tag{2.36}$$

that yields a value of K = 12.4 for a smooth UPVC stack that compares with empirical research results in this area, confirming the validity of the Chezy equation in this context (Figure 2.38).

The distance to terminal conditions may also be developed. If t′ is time then

$$\frac{dV}{dt'} = \frac{dV}{dz}\frac{dz}{dt'} = V\frac{dV}{dz}$$

and

$$\frac{dV}{dz} = \frac{1}{V}\frac{dV}{dt'} = g\frac{1}{V} - \frac{1}{2}f\frac{\pi D}{Q_w}V^3$$

so that

$$dz = \frac{1}{g}\frac{dV}{\left[1 - \left(\frac{f}{2g}\right)\left(\frac{\pi D}{Q_w}\right)\right]V^3}$$

Substituting for terminal velocity $V_t = \left(\frac{2g}{f}\frac{Q_w}{\pi D}\right)^{1/3}$ yields

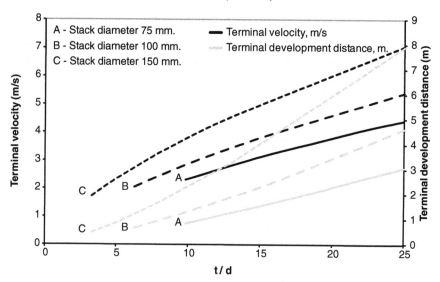

Figure 2.38 Terminal annular velocity and development distance in a vertical stack annular water flow for a range of stack diameters. Note data terminates when the annular flow reaches the allowable maximum area of 25% of the stack cross section.

$$dz = \frac{V_t^2}{g} \frac{\frac{V}{V_t} d\left(\frac{V}{V_t}\right)}{1-\left(\frac{V}{V_t}\right)} = \frac{V_t^2}{g} \frac{\theta d\theta}{(1-\theta^3)}$$

so that

$$\int_{z=0}^{z=Z} dz = \frac{V_t^2}{g} \int_{\theta=0}^{\theta=1} \frac{\theta d\theta}{(1-\theta^3)} \tag{2.37}$$

Integration yields the distance to terminal conditions; however the result will be infinite as the flow will approach terminal conditions asymptotically. A common approximation in such cases is to set terminal conditions at $\theta = 0.99°$ so that the vertical distance to achieve terminal conditions approximates to

$$Z = 01.59 V_t^2 \tag{2.38}$$

values included in Figures 2.33 and 2.34. It will be appreciated that rougher wall surfaces decrease both terminal velocity and development distance. Terminal thickness will increase as terminal velocity decreases with increasing wall roughness.

The falling water column will entrain an airflow in the stack core which in return results in a pressure drop as the air is drawn through any dry lengths of stack. Rapid changes in annular water downflow will also propagate low-amplitude air pressure transients to communicate the change in prevailing flow conditions. These transients are an essential component of the regime within building drainage networks and have been treated in detail elsewhere (Swaffield 2010).

2.9 Full bore flows in building drainage networks

Full bore flow will occur in building drainage networks following a system surcharge due to under sizing of the network or unforeseen increases in the system usage. These case are dealt with separately within unsteady flow models. In siphonic rainwater systems the design objective is to establish full bore flows in order to develop maximum roof discharges. In many cases the design methodology for siphonic systems remains based on steady flow conditions, and therefore it is appropriate to review briefly the relevant full bore flow equations of continuity, momentum and energy that are often applied.

In full bore flow the mass continuity equation remains

$$\sum \rho u A = \sum \rho Q = 0 \tag{2.39}$$

at all junctions and across any chosen control volume, effectively equation 2.5 as the flow density is constant.

The Steady Flow Energy Equation, equating potential, pressure and kinetic energy as well as the transfer of energy to the fluid internal energy through both frictional and separation losses, may be expressed across any control volume, Figure 2.39, as

$$\left(p+\rho gZ+\frac{1}{2}\rho u^2\right)_1 = \left(p+\rho gZ+\frac{1}{2}\rho u^2\right)_2 + \Delta p_{friction\ 1-2}$$
$$+ \Delta p_{separation\ 1-2} \tag{2.40}$$

$$\left(p+\rho gZ+\frac{1}{2}\rho\left(\frac{Q}{A}\right)^2\right)_1 = \left(p+\rho gZ+\frac{1}{2}\rho\left(\frac{Q}{A}\right)^2\right)_2 + \Delta p_{friction\ 1-2}$$
$$+ \Delta p_{separation\ 1-2} \tag{2.41}$$

The separation losses encountered at pipe bends, fittings and at entry and exit may be expressed in terms of the flow kinetic energy and an experimentally derived coefficient

$$\Delta p_{separation} = \frac{1}{2}\rho u^2 = \frac{1}{2}\rho\left(\frac{Q}{A}\right)^2 \tag{2.42}$$

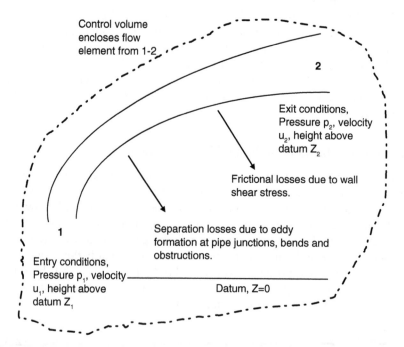

Figure 2.39 Development of the Steady Flow Energy Equation.

The frictional loss may be determined by consideration of the momentum equation applied across a short incremental length of conduit, Figure 2.40. Note that as the flow is steady the acceleration term is zero, so that

$$p_1 A - p_2 A - \tau_0 P \Delta x = \rho A (u_2 - u_1) = 0$$

and

$$\Delta p = \rho \frac{P}{A} \tau_0 \Delta x = \frac{1}{2} \rho \frac{P}{A} f u^2 \Delta x$$

when τ_0 is replaced by the friction factor and kinetic energy term as previously in the derivation of Chezy's equation.

Thus the frictional loss over a length L of conduit may be expressed as

$$\Delta p = \frac{\rho}{2m} f L u^2 = \frac{4 \rho f L}{2D} \left(\frac{Q}{A} \right)^2 \tag{2.43}$$

known as Darcy's Equation for full bore flow frictional losses. Note that the value of f may be determined from the Colebrook-White equation when $Re = \rho u D / \mu$ and $m = D/4$ for circular–cross section conduits, equation 2.44. It will also be seen that the Darcy and Chezy equations have identical formats.

$$\frac{1}{\sqrt{f}} = -4 \log_{10} \left(\frac{k}{3.71D} + \frac{1.26}{Re \sqrt{f}} \right) \tag{2.44}$$

Equations 2.39 to 2.43 will be used in the later treatment of siphonic rainwater systems where the design objective is to establish full bore flow.

2.10 Concluding remarks

The free surface flow equations developed, together with the frictional representations and the methodology for determining gradually varied flow depth

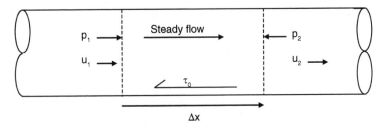

Figure 2.40 Development of Darcy's Equation, the full bore steady flow frictional relationship in a constant cross section conduit.

profiles and the location of hydraulic jumps will be essential in defining initial conditions in a building drainage network under steady flow conditions. However it is already clear, from the treatment of wave attenuation, for example, that the conditions within drainage networks are not steady but depend upon the random discharges of appliances to the network and the modification of those discharges due to their passage through the system. Discharge waves will attenuate, and the location of hydraulic jumps will oscillate dependent upon the changes in the approach flow or backflow from downstream junctions.

Hunter[5] (1924, 1940), working at the then National Bureau of Standards (NBS) in Washington, DC (now NIST), established much of the fundamental understanding of drainage flow conditions and was instrumental in developing early codes that still influence drainage design 70 years later. However it is clear that Hunter understood the limitations of his work when, in his 1940 paper, he stated that:

> the conventional pipe formulae apply to the irregular and intermittent flows that occur in plumbing systems only during that time (usually very short) and in that section of pipe in which the variable factors involved (velocity or volume rate of flow or hydraulic gradient and hydraulic radius) are constant.

This implies that he recognised that the treatment of the unsteady flow conditions was necessary but that the necessary mathematical simulation tools were not available. The extension of the understanding of free surface flows provided by Hunter would have to await the advent of accessible computing in the early 1960s. While the simulation of unsteady flows will be considered in detail in this text, this chapter has developed the fundamental equations of free surface flow in order to emphasise that the basic flow conditions within building drainage networks are amenable to analysis drawn from open channel fluid mechanics.

Notes

1 Antoine de Chézy (b. 1718, Châlons-sur-Marne; d. Oct. 5, 1798, Paris), French hydraulic engineer and author of a basic formula for calculating the velocity of a fluid stream derived from his studies in connection with the construction of French canals, notably in 1764 the difficult project of the Canal de Bourgogne, uniting the Seine and Rhône basins and first published in his paper 'Formula to find the uniform velocity that the water will have in a ditch or in a canal of which the slope is known'. This document resides in the collection of manuscripts in the library of the *École des Ponts et Chaussées*.

2 Robert Manning (b. 22 October 1816 Avincourt, Normandy; d. 9 December 1897) Irish civil engineer. During the 1860s, Manning carried out extensive scientific studies on aspects of rainfall, river volumes and water runoff. His experiments on open channel flow resulted in his classic 1889 paper 'On the flow of water in open

channels and pipes', followed by a further paper in 1895. His formula for open channel flow is one of the building blocks of the modern science of hydrology. Manning was President of the Institution of Civil Engineers of Ireland in 1877–1878.

3 Bisection method.

An iterative technique to solve for boundary conditions.

Set upper and lower limits to the solution, hu and hl,

Set acceptable level of f(0) to approach zero

A Set trial value as ht = 0.5(hu + hl),

Calculate function value f(xt); IF f(h) > 0, set hu = ht; IF f(h) < 0, set hl = ht

Return to A and recalculate ht and repeat until f(h) < f(0)

Solution x = xt

4 Simpson's Rule is credited to the English mathematician Thomas Simpson (1710–1761). For a function y = f(x) over an increment $x_0 - x_2$ where $x_1 = x_0 + h$ and $x_2 = x_0 + 2h$, the value of the integral may be expressed as $\int_{x_0}^{x_2} fxdx = \frac{h}{3}[f(x_0) + 4f(x_1) + f(x_2)]$.

5 Dr Roy B. Hunter (d. 1943). Recognised for his fundamental contributions to the understanding of building drainage system fluid mechanics and particularly for his introduction of the Hunter Fixture Unit approach to system pipe sizing. Fundamental reports from NBS in 1924 and 1940 remain a cornerstone of many national design codes. His major and unique technical contribution to plumbing code development was praised by Herbert Hoover as Secretary of Commerce. He established the binomial probability theory for simultaneous event analyses in water supply and drainage in BMS 65 (1940). Hunter understood that the fluid mechanics aspects of plumbing were poorly understood when he wrote in the 1924 code, '*Actual practice has been governed by opinions and guesswork, often involving needless costly precautions which many families could ill afford. The lack of generalized principles is responsible to a certain extent for the contradictory plumbing regulations in different localities*'. An opinion that could as easily be written 70 years after his untimely death in 1943.

3 Solution of the governing equations of fluid flow conditions in open channels and partially filled pipes

The defining unsteady flow equations of continuity and momentum were developed in the mid-nineteenth century by St Venant[1] (1870) as a pair of quasi-linear hyperbolic partial differential equations that may be solved by a variety of different numerical techniques.

This section will develop these governing equations before going on to detail the Method of Characteristics (MoC) solution technique. Alternative solution techniques will then be briefly discussed before developing the all-important boundary conditions.

3.1 Development of the general St Venant equations of continuity and momentum

The development of the St Venant equations that follows is general in that the pressure surge form of the equations will be required in a later discussion of siphonic rainwater system operation. However the derivations are general and provide the basis for free surface wave propagation simulation in both partially filled pipe flow and open rainwater gutters.

Figures 3.1 and 3.2 illustrate the unsteady flow conditions in an element of one-dimensional flow between two sections fixed in space δx apart, such that $\partial x / \partial t = 0$, within a conduit of general cross section that may or may not be flowing full. The conduit is assumed to be elastic and to be only subjected to small deformations due to the time-dependent pressure regime within the conduit. The conduit and the enclosed fluid do not move together as a rigid body – any change in conduit section length affects the conduit cross section through the action of the material Poisson's ratio. In order to retain a general solution the possibility of lateral inflow or extraction, q, is included in the derivation.

The continuity equation may be expressed (Figure 3.1) as

> Mass inflow – outflow = Rate of change of mass storage within the flow element

$$\rho Au - \left[\rho Au + \frac{\partial(\rho Au)\delta x}{\partial x} \right] + q\rho\delta x = \frac{\partial(\rho A\delta x)}{\partial t} \tag{3.1}$$

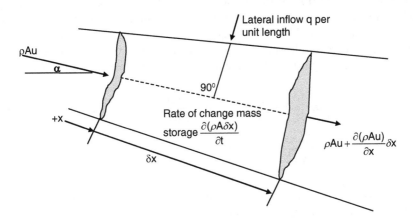

Figure 3.1 Derivation of the continuity equation for unsteady flow in a general conduit.

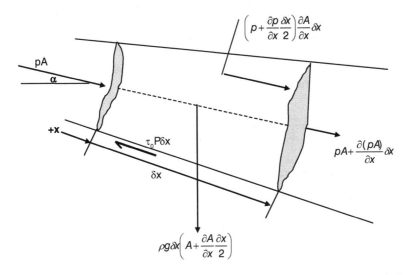

Figure 3.2 Derivation of the momentum equation for unsteady flow in a general conduit.

Expanding, collecting terms and dividing by $\rho A \partial s$ yields the continuity equation

$$\frac{\partial u}{\partial x} + \frac{1}{A}\left[\frac{\partial A}{\partial t} + u\frac{\partial A}{\partial x}\right] + \frac{1}{\rho}\left[\frac{\partial \rho}{\partial t} + u\frac{\partial \rho}{\partial x}\right] - \frac{q}{A} = 0 \qquad (3.2)$$

simplified as

Term 1c +Term 2c + Term 3c – Term 4c = 0

to allow a description of the influence of each term in turn and to identify which terms are appropriate etc. for any particular unsteady flow application (Douglas *et al.* 2005). Term 2c represents the effect of conduit cross-sectional change, while Term 3c represents the effect of density changes.

Similarly the equation of momentum may be expressed (Figure 3.2) as

$$pA + \left(p + \frac{\partial p}{\partial x}dx\right)\left(A + \frac{\partial A}{\partial x}dx\right) + \left(p + \frac{1}{2}\frac{\partial p}{\partial x}dx\right)\frac{\partial A}{\partial x}dx - \tau_0 Pdx - mg\sin\alpha$$

$$= \rho A dx \left(\frac{\partial u}{\partial t} + \frac{\partial u}{\partial x}\frac{dx}{dt}\right) + \rho q dx u$$

$$(3.3)$$

It is to be noted that the lateral inflow q is included in the momentum term, essential in the later treatment of lateral inflow to rainwater gutters but also applicable to land drain or defective drainage simulations, and that in the general case the conduit slope introduces a gravitational force. As the element is of general cross section, there is a force component derived from the pressure acting on the outer surfaces of the element. The wetted perimeter is referred to as P, and the shear stress acting between the fluid and the conduit wall is τ_0.

Expanding, neglecting second-order terms and dividing by ρA leads to the general momentum equation

$$\frac{1}{\rho}\frac{\partial p}{\partial x} + \left(\frac{\partial u}{\partial t} + \frac{\partial u}{\partial x}\frac{dx}{dt}\right) - g\sin\alpha + \frac{\tau_0 P}{\rho A} + \frac{qu}{A} = 0$$

$$(3.4)$$

where it is usual to write $\tau_0 = \frac{1}{2}\rho f u^2$

It is again convenient to represent this expression as

Term 1m + Term 2m − Term 3m + Term 4m + Term 5m = 0

Each term has its particular influence, e.g. Term 3m represents the effect of conduit slope, Term 4m represents the shear forces between the flow and the enclosing surface and Term 5m the lateral inflow contribution to flow momentum change.

Table 3.1 illustrates the necessity to include each of these terms in either the continuity or momentum equation dependent upon the unsteady flow regime to be considered (Douglas *et al.* 2005).

In the case of free surface flow the independent variables are flow depth and mean flow velocity. The density change Term 3c may be ignored, and for partially filled conduit flow without a lateral inflow the Term 4c may also be ignored. The area change Term 2c may be expressed (Figure 3.3) as

$$dA = Tdh$$

$$(3.5)$$

Table 3.1 Relevance of each term identified within the St Venant unsteady flow
equations of continuity and momentum

Unsteady flow regime	Variables	Continuity equation 3.2	Momentum equation 3.4
Traditional waterhammer, including closed conduit flows, siphonic rainwater systems	Pressure p, velocity u, distance x and time t.	Terms 1, 2, 3	Terms 1, 2, 3, 4
Free surface flows, including building drainage systems within or close to the building envelope	Depth h, velocity u, distance x and time t	Terms 1, 2 only	Terms 1, 2, 3, 4
Free surface flows where lateral inflow becomes important	Depth h, velocity u, distance x and time t	Terms 1, 2, 4 only	Terms 1, 2, 3, 4, 5
Low-amplitude air pressure transient applications, simplifications include no changes in conduit cross section, no lateral inflows or longitudinal duct extensions.	Wave speed c, velocity u, distance x and time t (Note pressure p is related to density and is determined from the wave speed prediction.)	Terms 1, 3 only	Terms 1, 2, 4 only

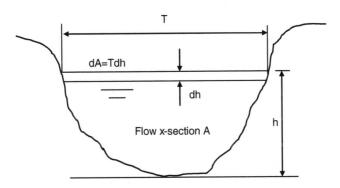

Figure 3.3 General free surface flow conduit properties.

The free surface wave propagation speed has already been shown to be

$$c = \sqrt{\frac{gA}{T}} \tag{3.6}$$

and so the continuity equation for free surface flow conditions, with or
without lateral inflow, becomes

$$c^2 \frac{\partial u}{\partial x} + g\left(\frac{\partial h}{\partial t} + u\frac{\partial h}{\partial x}\right) - \frac{c^2 q}{A} = 0 \tag{3.7}$$

Within the momentum equation 3.4 the frictional resistance Term 4m may be expressed via the Chezy equation, equation 2.11, and a suitable friction factor, preferably determined from the Colebrook-White expression, (Bridge and Swaffield 1983):

$$\text{Term } 4c = \left(\frac{1}{2}\rho f u^2 \frac{P}{\rho A}\right) = \left[\frac{1}{2}\frac{f}{m}\right]\left[\frac{2g}{f}Sm\right] = gS \tag{3.8}$$

where S is the friction slope, or the channel invert slope under steady Normal flow conditions.

The momentum equation 3.4 therefore becomes

$$g\frac{\partial h}{\partial x} + \left[\frac{\partial u}{\partial t} + u\frac{\partial u}{\partial x}\right] + g(S - S_0) + \frac{uq}{A} = 0 \tag{3.9}$$

where the lateral inflow term is retained for later reference and S_0 is the channel invert slope. (Note that in steady uniform flow the differential terms tend to zero and S equates to S_0.)

The advent of siphonic rainwater drainage systems has introduced full bore flow as the norm in the drainage network serving the roof-level outlets. In this case the propagation of system condition change will depend upon the propagation of pressure transients, and the appropriate St Venant equations of continuity and momentum have to include the pipe elasticity and the fluid density and Bulk Modulus, as the transient propagation velocity will depend on both pipe and fluid characteristics, and the treatment is identical to that for classical waterhammer (Swaffield and Boldy 1993; Swaffield 2010).

In this case the variables are pressure, p, velocity, u, distance, x, and time, t. In the continuity equation the density, cross-sectional area and longitudinal extension variation terms may be combined through the introduction of wave propagation velocity, c.

The density change Term 3c may be expressed as

$$\text{Term } 3c = \frac{1}{K}\left[\frac{\partial p}{\partial t} + \frac{\partial p}{\partial x}\frac{dx}{dt}\right] \tag{3.10}$$

where K is the fluid bulk modulus defined as $dp/K = d\rho/\rho$.

The cross-sectional area Term 2c may be expressed (Swaffield 2010) in terms of the axial and lateral stress imposed on the conduit wall by the passing pressure transient, the conduit material Young's Modulus of Elasticity, the conduit wall thickness and the degree of restraint applied to the conduit through its supports. (The effect of conduit restraint is normally included by introducing a constant C_1 whose value depends on the conditions imposed; e.g. $C_1 = 1$ if the conduit is fully restrained along its whole length.)

It may be shown that Term 2c becomes

$$\text{Term 2c} = \frac{D}{Ee}C_1\left[\frac{\partial p}{\partial t} + \frac{\partial p}{\partial x}\frac{dx}{dt}\right] \tag{3.11}$$

allowing a combination of Terms 2c and 3c and an identification of this term as representing wave propagation velocity in the fluid as modified by the presence of an elastic pipe wall.

It is convenient to write an equivalent bulk modulus for the combined air and water mixture compression and pipe wall distortion as

$$\frac{1}{K_{eff}} = \frac{1-y}{K_w} + \frac{y}{K_a} + \frac{DC_1}{Ee}$$

$$K_{eff} = 1/\left[\frac{1-y}{K_w} + \frac{y}{K_a} + \frac{DC_1}{Ee}\right] \tag{3.12}$$

where y is the undissolved air content of the water flow and the wave propagation velocity in a fluid as

$$c = \sqrt{\frac{K_{\varepsilon eff}}{\rho_{eff}}} \tag{3.13}$$

where y is the air proportion by volume at NTP so that the bulk modulus and density terms are seen to incorporate the effect of wall distortion and any entrained free air. The effective density is expressed as

$$\rho_{eff} = y\rho_a + (1-y)\rho_w \tag{3.14}$$

The presence of free air in the rainwater flow due to entrainment at the gutter outlet is high, and the full expression for wave speed in an entrained air and water mixture in an elastic pipe is required:

$$c = \sqrt{\frac{K_{eff}}{\rho_{\varepsilon\phi\phi}}} = \sqrt{1/\left[\left\{\frac{1-y}{K_w} + \frac{y}{K_a} + \frac{DC_1}{Ee}\right\}\left\{\frac{1}{y\rho_a + (1-y)\rho_w}\right\}\right]} \tag{3.15}$$

where suffixes $_a$ and $_w$ refer to the entrained air and water respectively.

The Young's Modulus for pipe materials commonly found in building drainage systems and the Bulk Modulus of water at relevant temperatures are illustrated by Tables 3.2 and 3.3, while typical wave speed values in full bore flow with and without free air content are illustrated by Figure 3.4.

The continuity equation therefore may be written as

$$\rho c^2 \frac{\partial u}{\partial x} + \frac{\partial p}{\partial t} + u \frac{\partial p}{\partial x} = 0 \tag{3.16}$$

Referring to the equation of motion, equation 3.4 it may be seen that for a full bore flow case the frictional resistance Term 4m may be expressed in terms of the Darcy equation and a suitable friction factor, preferably determined from the Colebrook-White expression, equation 2.44:

$$\text{Term 4m} = \frac{fu^2}{m} \tag{3.17}$$

where m is the hydraulic mean depth or hydraulic radius, equal to D/4 for full bore flow in circular–cross section conduits.

Table 3.2 Values of Young's Modulus for possible pipe materials

Material	Young's Modulus (10^{-9} N/m^2)	Poisson's Ratio
Aluminium	70.0	0.3
Cast iron	80.0–100.0	0.25
Copper	107.0–130.0	0.1–0.3
Glass	68.0	0.24
GRP	50.0	0.35
Polythene	3.1	—
PTFE plastic	0.35	—
PVC plastic	2.4–2.8	—
Steel	200.0–214.0	0.3
Titanium	103.4	0.34

Table 3.3 Values of water Bulk Modulus and density

Water at temperature, °C	Bulk modulus (10^{-8} N/m^2)	Density (kg/m^3)
0	20.5	1000.0
20	20.5	998.0
80	20.5	972.0
Sea water at 10 °C	22.0	1026.0

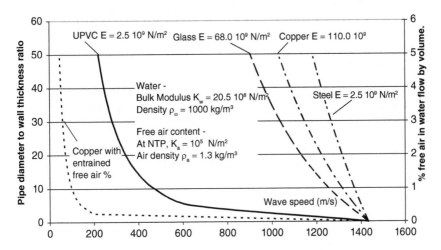

Figure 3.4 Wave speed in full bore water flow in a range of pipe materials to demonstrate impact of fluid and pipe properties. Influence of free air is demonstrated for % free air content in a copper pipe with a D/e ratio of 8. (Note wave speed as x-axis to allow comparison of air free and entrained air cases.)

As the lateral inflow Term 5m is naturally excluded, the equation of motion, equation 3.4, thus becomes

$$\frac{1}{\rho}\frac{\partial p}{\partial x} + \left[\frac{\partial u}{\partial t} + u\frac{\partial u}{\partial x}\right] - g\sin\alpha + \frac{fu|u|}{2m} = 0 \tag{3.18}$$

3.2 Application of the Method of Characteristics (MoC) to model the unsteady flow regimes met in building drainage and rainwater drainage networks

In each of the preceding cases, the St Venant equations are recognised as a pair of quasi-linear hyperbolic partial differential equations that may be solved numerically provided a scheme to transform the partial into total derivatives can be identified. The Method of Characteristics (MoC) was first used by Rieman in 1860 to study sound wave propagation and later by Massau in 1900 to consider free surface wave motion. The first application to pressure surge analysis is due to Lamoen (1947), Gray (1953) and Ezekial and Paynter (1957); however none of these contributions made reference to the impact of access to fast digital computing. Lister (1960) established the application of MoC to solve the waterhammer equations in her seminal paper that provided the underpinning for the application of the methodology discussed here to waterhammer analysis by Streeter and Lai (1962) in the USA and in the UK by Fox (1968). While Streeter considered applications to open channel wave propagation, the first specific application of MoC to building drainage network analysis was by Swaffield (1980a) at NBS, Washington, DC.

3.2.1 Transformation of governing equations to characteristic equations

The equations of continuity and motion in two dependent variables, Λ_1 and Λ_2, and two independent variables, x and t, may be written as:

$$L_1 = \alpha_c \frac{\partial \Lambda_1}{\partial x} + \beta_c \left[\frac{\partial \Lambda_2}{\partial t} + V_1 \frac{\partial \Lambda_2}{\partial x} \right] + \gamma_c = 0 \tag{3.19}$$

$$L_2 = \alpha_m \frac{\partial \Lambda_2}{\partial x} + \beta_m \left[\frac{\partial \Lambda_1}{\partial t} + V_1 \frac{\partial \Lambda_1}{\partial x} \right] + \gamma_m = 0 \tag{3.20}$$

where the values of the coefficients α, β, χ for each case are listed in Table 3.3.

Combining equations 3.19 and 3.20 as $L_1 + \lambda L_2 = 0$, where λ is an arbitrary constant, yields:

$$\lambda \beta_m \left[\frac{\partial \Lambda_1}{\partial t} + \frac{\partial \Lambda_1}{\partial x} \left(\Lambda_1 + \frac{\alpha_c}{\lambda \beta_m} \right) \right] + \beta_c \left[\frac{\partial \Lambda_2}{\partial t} + \frac{\partial \Lambda_2}{\partial x} \left\{ \Lambda_1 + \frac{\lambda \alpha_m}{\beta_c} \right\} \right]$$
$$+ \gamma_c + \lambda \gamma_m = 0 \tag{3.21}$$

By inspection it follows that the combined expression, equation 3.21, may be expressed as a total differential equation of the form

$$\beta_c \frac{d\Lambda_2}{dt} + \lambda \beta_m \frac{d\Lambda_1}{dt} + (\gamma_c + \lambda \gamma_m) = 0 \tag{3.22}$$

provided that

$$\frac{dx}{dt} = \Lambda_1 + \frac{\alpha_c}{\lambda \beta_m} = \Lambda_1 + \frac{\lambda \alpha_m}{\beta_c}$$

hence

$$\lambda = \pm \sqrt{\frac{\alpha_c \beta_c}{\alpha_m \beta_m}} \tag{3.23}$$

and

$$\frac{dx}{dt} = \Lambda_1 \pm \sqrt{\frac{\alpha_c \alpha_m}{\beta_c \beta_m}} \tag{3.24}$$

The combined total differential equation may thus be written as

$$\frac{d\Lambda_1}{dt} \pm C_2 \frac{d\Lambda_2}{dt} + C_3 = 0 \tag{3.25}$$

provided that

$$\frac{dx}{dt} = \Lambda_1 \pm c \tag{3.26}$$

This relationship between time step, internodal length and the flow and wave propagation velocities is central to the Method of Characteristics solution of the St Venant equations. It is known as the Courant Criterion and applies to all the cases of transient propagation to be covered in this text. Adherence to the Courant Criterion also dictates the time step for any network, as all constituent pipes must have the same time step to allow continuity at junction boundaries. It follows therefore that the time step depends upon the largest combined value of flow and wave propagation velocities found within the network, and this in turn leads to some of the computational difficulties to be discussed later, in particular the possibility of rounding errors or backwater profile collapse due to excessive interpolation brought about by very different wave and flow velocities within the same network.

Referring to Figure 3.5 and Table 3.5 it will be seen that the St Venant equations may be represented by two first-order finite difference equations, based on equation 3.25, known respectively as C⁺ and C⁻ characteristics linking known conditions at time t at points R and S to conditions at P at one time step in the future. These may be expressed as:

C⁺ characteristic:

$$\Lambda_1^P - \Lambda_1^R + C_2 \left[\Lambda_2^P - \Lambda_2^R \right] + C_3 \, \Delta t = 0 \tag{3.27}$$

provided that

$$x^P - x^R = \left[\Lambda_1^R + c^R \right] \Delta t \tag{3.28}$$

and the C⁻ characteristic:

$$\Lambda_1^P - \Lambda_1^S - C_2 \left[\Lambda_2^P - \Lambda_2^S \right] + C_3 \, \Delta t = 0 \tag{3.29}$$

provided that

$$x^P - x^S = \left[\Lambda_1^S - c^S \right] \Delta t \tag{3.30}$$

where the coefficients C_2 and C_3 are given by Table 3.4 and superscripts P, R, S refer to Figure 3.5 which illustrates a number of points inherent in the application of the Method of Characteristics to this family of equations, considered in more detail in the following paragraphs. Note that in the special case of supercritical free surface flow the 'downstream' point S becomes upstream point S'.

For each flow case then the characteristic equations reduce to a general format that may be expressed as

$$\Lambda_1^P = \left[\Lambda_1^R + C_2 \Lambda_2^R - C_3 \, \Delta t \right] - C_2 \Lambda_2^P = K1 - K2 \Lambda_2^P = 0 \tag{3.31}$$

provided that

$$x^P - x^R = \left[\Lambda_1^R + c^R \right] \Delta t \tag{3.32}$$

$$\Lambda_1^P = \left[\Lambda_1^S - C_2 \Lambda_2^S - C_3 \Delta t \right] + C_2 \Lambda_2^P = K3 + K4 \Lambda_2^P = 0 \tag{3.33}$$

provided that

$$x^P - x^S = \left[\Lambda_1^S - c^S \right] \Delta t \tag{3.34}$$

where Table 3.4 defines the values of the coefficients K1–4 based on known conditions at time t.

The characteristic equations developed and defined by Table 3.6 allow the propagation of both free surface unsteady flow and the treatment of pressure transients in full bore siphonic rainwater systems. In order to develop the simulation techniques to be discussed in this text it is now necessary to outline the development of initial conditions necessary at the start of any simulation and the interpolation techniques that allow the known conditions at time t, at the base of each characteristic (Figure 3.5) to be determined.

Table 3.4 Identification of dependent variables and coefficients in the equations of continuity and motion developed for unsteady building drainage applications

Flow condition	Dependent variables		Continuity equation			Momentum equation			Venant multiplier $\lambda = \sqrt{\alpha_c \beta_c / \alpha_m \beta_m}$	Courant Criterion dx/dt
	Λ_1	Λ_2	α	β	γ	α	β	γ		
Partially filled pipe flow	u	h	c^2	g	0	g	1	$(g(S-S_0))$	$\pm c$	$u\pm c$
Open roof gutter flow and lateral inflow land drain	u	h	c^2	g	$(-c^2 q/A)$	g	1	$(g(S-S_0) + qu/A)$	$\pm c$	$u\pm c$
Siphonic rainwater full bore flow	u	p	ρc^2	1	0	$1/\rho$	1	$(-g\sin\alpha + fu|u|/2m)$	$\pm pc$	$u\pm c$

Table 3.5 Identification of the coefficients in the finite difference equations applicable to the building drainage applications considered

Flow condition	$C_2 = \beta_c / \lambda\beta_m$	$C_3 = (\gamma_c + \lambda\gamma_m)/\lambda\beta_m$	Wave speed c
Partially filled pipe flow	g/c	$g(S-S0)$	$c = \sqrt{\dfrac{gA}{T}}$
Open roof gutter flow and lateral inflow land drain	g/c	$g(S-S0) + q(u\pm c)/A$	$c = \sqrt{\dfrac{gA}{T}}$
Siphonic rainwater full bore flow	$1/\rho c$	$-g\sin\alpha + fu|u|/2m$	$c = \sqrt{\dfrac{K_{\text{eff}}}{\rho_{\varepsilon\phi\phi}}} = \sqrt{\dfrac{(1-y)/K_w + y/K_a + DC_1/Ee}{(1-y)\rho_w + y\rho_a}}$

Table 3.6 C⁺ and C⁻ characteristic equations for each of the free surface and siphonic system full bore flow cases considered

Flow condition	Λ_1	0_2	K1	K3	K2 and K4
Free surface partially filled pipe flow case	uP	hP	uR + (g/c)hR – g(S – S0)Δt	uR – (g/c)hR – g(S – S0)Δt	g/c
Gutter or lateral flow land drain	uP	hP	uR + (g/c)hR – (g(S – S0) + q(u – c)/A Δt	uR – (g/c)hR – (g(S – S0) + q(u + c)/A) Δt	g/c
Siphonic full bore flow case	uP	pP	uR + (1/ρc)pR – (gsinα + fu|u|/2m) Δt	uR – (1/ρc) pR – (gsinα + fu|u|/2m) Δt	1/ρc

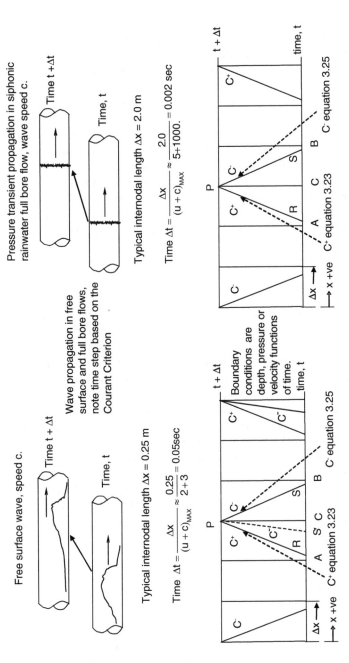

Figure 3.5 Method of Characteristics representation of unsteady free surface and full bore conditions. Note that in the case of unsteady free surface supercritical flow both characteristics slope downstream.

3.2.2 Initial conditions

Initial conditions at time zero must be established along the whole length of each conduit making up the network in order to determine the initial values of the velocity, u, depth, h, or pressure, p, variables and the location of each of the base points, R, S and S' for each internodal reach. For airflow in building drainage systems or for full bore flow in siphonic systems or during drain surcharge, these conditions may be zero flow velocity and some system pressure, possibly atmospheric. In the case of free surface fluid flows it is necessary to define a 'trickle' flow with flow depth and velocity calculated at each node. Where a free surface flow is subcritical it will be necessary to define any backwater profile and the location of any hydraulic jumps in the case of supercritical flows interacting with system junctions or obstructions by employing the equations of free surface gradually varied flow introduced in Chapter 2. The inclusion of an initial 'trickle flow', possibly of the order of 0.1 litres/second, is defensible as in practice drainage networks will normally have residual flows for a considerable period following appliance discharge due to wave attenuation.

3.2.3 Simulation methodology

Figures 3.6 a–d illustrate a system of characteristic equations superimposed on illustrative pipe and conduit sections to demonstrate the mechanism of the MoC simulation. The values of wave speed and mean flow velocity will vary considerably between the four examples shown, so the time steps and internodal reach lengths will vary in order to satisfy the Courant Criterion.

For full bore flow cases in siphonic rainwater systems the pressure transient propagation wave speed in the water will be affected by the elasticity of the pipe wall, the wall-to-diameter ratio and the free air content in the water flow entrained through the gutter outlet – despite the outlet being designed to restrict air entrainment, this will occur during the establishment of full bore flow siphonic running conditions. For a copper pipe with a diameter-to-wall thickness ratio of 8 the wave speed will vary from 1300 m/s with no air entrainment to 130 m/s with 1% free air content. If the mean water flow velocity is taken as 5 m/s then the time step to satisfy the Courant Criterion,

$$\Delta t = \frac{\Delta x}{(u+c)_{max}} \tag{3.35}$$

will therefore vary from 1/1300 to 1/140 seconds. However as c >> u the C⁻ characteristic will continue to slope upstream. In addition a C⁻ characteristic will exist at pipe entry to allow solution with the pipe entry flow boundary condition (Figure 3.6a).

(a) *Full* bore flow transient simulation as found in siphonic rainwater systems, wave speed typically 100–1000 m/s. C^+ characteristic slopes downstream, C^- characteristic slopes upstream.

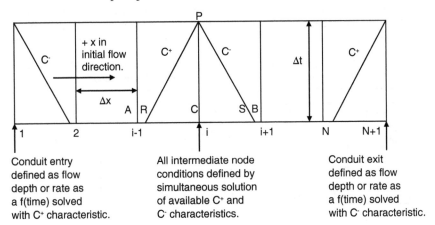

Conduit entry defined as flow depth or rate as a f(time) solved with C^+ characteristic.

All intermediate node conditions defined by simultaneous solution of available C^+ and C^- characteristics.

Conduit exit defined as flow depth or rate as a f(time) solved with C^- characteristic.

(b) *Subcritical* free surface flow as found in shallow gradient branch drains or roof gutters. C^+ characteristic slopes downstream, C^- characteristic slopes upstream until the flow mean velocity, dependent upon channel slope, hydraulic mean depth, roughness and applied flowrate, approaches the flow wave speed, dependent on flow cross sectional area and surface width only.

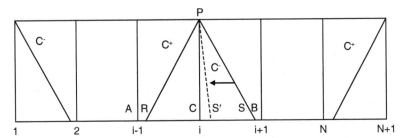

(c) *Supercritical* free surface flow as found in steep gradient branch drains. C^+ characteristic slopes downstream; C^- characteristic also slopes downstream. Note no boundary condition solution necessary at channel free outfall as both characteristics present. At entry no characteristic exists to solve with an imposed upstream boundary.

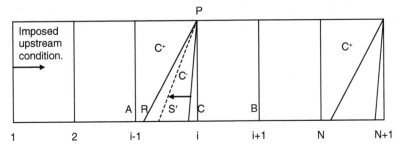

(d) *Supercritical* free surface flow forced to modify to subcritical through a hydraulic jump, as found in steep gradient branch drains terminated by a junction or obstruction. C^+ characteristic slopes downstream, C^- characteristic also slopes downstream in the supercritical flow section. Transition occurs at the jump which may not align with a internodal reach boundary introducing interpolation errors.

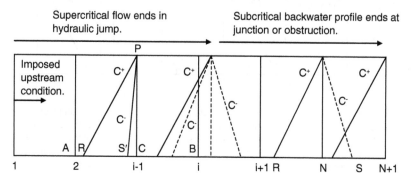

Figure 3.6 a–d. Characteristic equations available in full bore transient simulation, partially filled subcritical and supercritical conduit flows and mixed regime supercritical and subcritical flows across a hydraulic jump upstream of a junction or obstruction.

In the case of subcritical free surface wave propagation the wave speed will, in a 100 mm diameter drain, be less than 3 m/s, dependent only on the flow depth and surface width. The mean flow velocity however may be more or less than this dependent upon drain diameter, roughness and slope – determined via the Manning Equation. In subcritical flows u < c, and the C+ and C– characteristics will slope downstream and upstream respectively, as shown in Figure 3.6 b. It will be appreciated from Figure 3.6 b that as u approaches c the C– characteristic will approach the vertical.

In supercritical flows both characteristics slope downstream, as shown in Figure 3.6 c, which implies that information cannot be propagated upstream. For this reason supercritical flow leaving a conduit via a free outfall 'remains unaware' of the pipe termination, the exit condition being defined by the simultaneous solution of the available characteristics as shown. Similarly there is no available C⁻ characteristic available at channel entry, the entry condition being imposed – for example the presence of a Critical depth at entry to a supercritical flow from a junction or downstream of an obstruction. It will be appreciated that in order to impose entry depth and velocity values it will be necessary to solve the local flowrate vs. time profile with the Critical depth equation.

Figure 3.6 d illustrates the mixed flow regime upstream of a junction or obstruction. The junction boundary will impose a local flow depth greater than flow Critical depth, and a hydraulic jump will be formed upstream linked to the junction depth by a backwater profile. Upstream of the jump the approach flow will be supercritical. This introduces the necessity to express the hydraulic jump in terms of a boundary condition capable of movement in both directions in the channel flow depending on approach flow and junction conditions.

The application of the Method of Characteristics to the solution of the St Venant equations of continuity and momentum defining transient propagation by Lister (1960) enabled a paradigm shift in the simulation of pressure transient propagation. Early work by Streeter and Fox employed Lister's approach which results in a surprisingly simplistic methodology that effectively has only two major components. Future conditions at all internal nodes Δx apart are predicted by simultaneous solution of the available C⁺ and C⁻ characteristics, whether in full bore flow applications or in either subcritical or supercritical free surface flows. Future conditions at pipe entry or termination are predicted by solving whichever characteristic is available with a suitable boundary condition at that location.

As already mentioned, entry to a supercritical flow section will require the imposition of flow depth and velocity values by solution of the local flow–time profile with some assumption as to the flow regime – usually Critical depth at pipe entry from a junction or downstream of an obstruction of slope defect.

This definition of the methodology is of course oversimplified, as the simulation requires careful choice of suitable boundary conditions and in all cases knowledge of the flow conditions at the base of each characteristic at the start of the time step, locations R, S and S' in Figures 3.5 and 3.6. This implies the introduction of some form of interpolation technique as conditions are only known at the nodal points A, B and C. This requirement is the most important element in successful applications of MoC and the cause of the greatest computational difficulties and potential errors.

Figures 3.6 a–d illustrate a form of the solution technique known as specified time intervals where the time step is set by reference to the largest combination of velocity and wave speed, equation 3.35.

3.2.4 Interpolation techniques

The rectangular grid is based on fixed and chosen internodal lengths and a time step calculation that ensures that points R and S always fall within an internodal length upstream or downstream of the calculation node, defined at P. It will be appreciated that for a network the time step must be constant for all the interlinked pipes, and this may lead to the necessity to vary the internodal length between pipes. As Figure 3.6 presents the specified time interval approach, interpolations to determine conditions at R and S are required along the distance axis at time t. This may lead to rounding errors in some cases as discussed further later.

It will be seen from Figure 3.6 that halving the internodal length will quadruple the number of calculation steps to complete any given simulation time, a consideration when the computing capacity available was limited.

Variable wave speed along the length of a pipe, duct or channel leads to different slopes for the C^+ and C^- characteristics. Assuming that initial conditions are known at all nodes at time t, then it is necessary to determine the conditions at R and S at time t in order to apply the characteristics solution already developed. The most common solution to this situation is to interpolate linearly between conditions at A and C to obtain conditions at R and similarly between C and B to obtain conditions at S. (Note here that if c < u, i.e. the supercritical free surface or supersonic flow condition, then both R and S lie between A and C in Figure 3.6; in this case S is usually referred to as S'.)

It is worth noting that interpolation implies that a pressure transient or surface wave arriving at A or B at time t determines conditions at R, S or S' at that time. This effectively increases the speed of propagation of the transient and also decreases the rate of change of pressure or velocity that it imparts to the flow it passes through. Both effects lead to a rounding in the predicted transient.

Taking the free surface flow case as representative of interpolation techniques where the velocity, u, and depth, h, stand for the Λ_1 and Λ_2 variables in Tables 3.4 and 3.6 and referring to Figure 3.6 as the general case where both c and u vary along the conduit and also where the possibility that u > c is allowed to exist so that both C^+ and C^- characteristics slope downstream, a series of equations may be presented linking conditions at R and S (or S') to conditions at the nodes A, C and B.

For the C^+ characteristic passing through R and linking conditions at R at time t to conditions at P at time t+Δt, consideration of the velocity variation yields

$$\frac{u_C - u_R}{u_C - u_A} = \frac{x_C - x_R}{\Delta x} = (u_R + c_R)\frac{\Delta t}{\Delta x}$$

since

$$x_C = x_P$$

and

$$\Delta x = x_C - x_A$$

Similarly the wave speed terms yield

$$\frac{c_C - c_R}{c_C - c_A} = \frac{x_C - x_R}{\Delta x} = (u_R + c_R)\frac{\Delta t}{\Delta x}$$

Simultaneous solution of these equations results in a series of interpolation relationships that allow the determination of base conditions at point R

$$u_R = \frac{u_C + \theta(u_A c_C - u_C c_A)}{1 + \theta(u_C - u_A + c_C - c_A)} \tag{3.36}$$

$$c_R = \frac{c_C + \theta u_R(c_A - c_C)}{(1 + \theta(c_C - c_A))} \tag{3.37}$$

where

$$\theta = \frac{\Delta t}{\Delta x}$$

and for the depth or pressure terms

$$h_R = h_C - (h_C - h_A)\theta(u_R + c_R) \tag{3.38}$$

$$p_R = p_C - (p_C - p_A)\theta(u_R + c_R) \tag{3.39}$$

A number of points need to be stressed concerning the above equations:

1 If the wave speed is a constant, but still comparable to the local flow velocity, these equations will yield the interpolated values of flow velocity u with no modification.
2 In cases where the velocity is negligible with respect to a constant wave speed, so that

$$c_P = c_A = c_B$$

these interpolation equations may be simplified as

$$\frac{u_C - u_R}{u_C - u_A} = \theta c_A$$

thus

$$u_R = u_C - \theta c_A(u_C - u_A)$$

and pressure, p, or depth, h, is given by

$$p_R = p_C - \theta c_A(p_C - p_A)$$

$$h_R = h_C - \theta c_A(h_C - h_A)$$

Under subcritical flow conditions the C⁻ characteristic slopes downstream and the base conditions are found at point S in Figure 3.6. By a similar series of substitutions to those above the following expressions may be derived

$$u_S = \frac{u_C - \theta(u_C c_B - u_B c_C)}{1 - \theta(u_C - u_B - c_C + c_B)} \tag{3.40}$$

$$c_S = \frac{c_C + \theta u_S(c_C - c_B)}{(1 + \theta(c_C - c_B))} \tag{3.41}$$

and for the pressure or head terms

$$p_S = p_C + (p_C - p_B)\theta(u_S - c_S) \tag{3.42}$$

$$h_S = h_C + (h_C - h_B)\theta(u_S - c_S) \tag{3.43}$$

If c >> u and is assumed to be constant, then these equations reduce as before to

$$\frac{u_C - u_S}{u_C - u_B} = \theta c_B$$

thus

$$u_S = u_C + \theta c_B(u_B - u_C)$$

and pressure, p, or depth, h, is given by

$$p_S = p_C + \theta c_B(p_B - p_C)$$
$$h_S = h_C + \theta c_B(h_B - h_C)$$

If the local flow velocity exceeds the wave speed, a supercritical flow regime, then the C⁻ characteristic also slopes downstream and the required base point, S′, is to be found upstream of point P. This condition would not be expected in the analysis of pressure transients; however it is the norm in many free surface flow applications of the Method of Characteristics. In this case the interpolation equations are derived in the same manner as set out above, resulting in the following expressions:

$$u_{S'} = \frac{u_C - \theta(u_A c_C + u_C c_A)}{1 + \theta(u_C - u_A + c_A - c_C)} \tag{3.44}$$

$$c_{S'} = \frac{c_C + \theta u_{S'}(c_A - c_C)}{(1 + \theta(c_A - c_C))} \tag{3.45}$$

and for the pressure or head terms

$$p_{S'} = p_C - (p_C - p_A)\theta(u_{S'} - c_{S'}) \tag{3.46}$$

$$h_{S'} = h_C - (h_C - h_A)\theta(u_{S'} - c_{S'}) \tag{3.47}$$

Lister (1960) advocated a form of linear interpolation combined with an iterative approach. Initial values for the characteristic slopes in equations 3.32 and 3.34 were based on values at node C, i.e. node P one time step earlier. Interpolated values at R and S were then determined from equations 3.36 to 3.47 dependent upon flow regime. Values at node P were then determined by simultaneous solution of equations 3.31 and 3.33. The process was repeated with values at P being used to determine the characteristic slopes until these predictions converged. Generally this iterative approach was not followed in the early papers by Streeter and Lai (1962) or Fox (1968), the direct single pass technique developed above being favoured.

While linear interpolation proved sufficiently accurate for the application of Lister's solutions to the traditional pressure surge/waterhammer applications and to the later application to low-amplitude air pressure transient simulation in building drainage vent systems (Swaffield 2010), the sensitivity of the solution to errors in approximating free surface flow depths between known nodal values has resulted in serious computational difficulties. Figure 3.7 illustrates the potential error within the drawdown profile close to a free discharge in a subcritical flow.

The error in interpolation ignores the natural curvature of the free water surface and also implies that disturbances arriving at the upstream and downstream nodes at time t affect the interpolated value at that time – effectively smoothing the wave front and increasing its propagation velocity. The most evident example of this error involves the progressive collapse of the drawdown profile upstream of a free outfall or the collapse of the backwater profile upstream of a junction or obstruction that causes an upstream hydraulic jump.

Several approaches to limit the effect of interpolation rounding errors have been reported. Goldberg and Wylie (1983) introduced a 'time line' interpolation (Figure 3.8) where interpolations were undertaken at each node over previous time steps in order to avoid implying that transient effects affected values ahead of the wave position at any time. Time line interpolation is highly effective where there is a period of steady flow, as

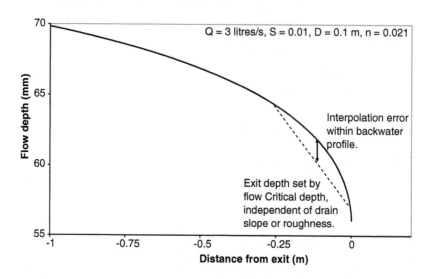

Figure 3.7 Linear interpolation errors demonstrated within the backwater profile upstream of a subcritical free outfall.

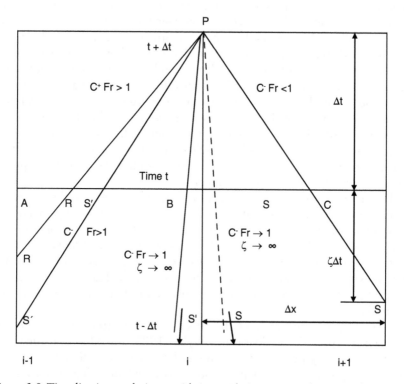

Figure 3.8 Time line interpolation avoids interpolation errors inherent in the linear interpolation scheme. Note storage becomes impractical as the C⁻ characteristic approaches the vertical.

the values of velocity and depth at each node remain constant and hence interpolation becomes trivial. However the main problem with time line interpolation arises if the characteristic slopes approach the vertical – i.e. at the transition from sub to supercritical flows. The back storage capacity requirement then tends to infinity.

The effect of rounding errors arising from linear interpolation is clearly demonstrated in a full bore flow transient simulation of a siphonic rainwater system where the roof outlet becomes suddenly and completely blocked. This generates a negative transient downstream of the roof gutter outlet as shown in Figure 3.9. In full bore steady flow, interpolation errors are not an issue as the velocity is constant in the flow direction, and the pressure loss due to friction is also constant for steady uniform flow in a single conduit. However, once the flow becomes unsteady, the interpolation process introduces rounding errors that reduce peak excursions and timing errors as the effect of the transient is assumed ahead of its actual position. Comparative traces are shown with the time step determined by the Courant Criterion and a 50% time step where the characteristic base values at R and S are determined by linear interpolation. The time step is defined as

$$dt = \frac{dx}{TFAC(u+c)_{max}} \tag{3.48}$$

where the TFAC = 1 represents the Courant Criterion time step and TFAC > 1 indicates interpolation at successively smaller time steps.

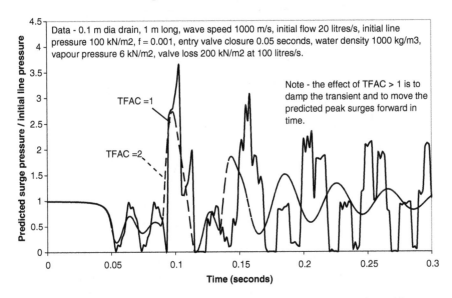

Figure 3.9 Impact of TFAC > 1 on the predicted surge downstream of a sudden stoppage in a siphonic rainwater system.

As shown in Figure 3.10 the level of interpolation rounding error will increase as the time step is progressively reduced below the Courant Criterion value, i.e. as TFAC increases above 1. It will be seen (Figure 3.9) that the peak pressure predictions are substantially reduced and the time of occurrence of maximum surge pressure is reduced on each successive peak – results consistent with a rounding error and an assumption that transient changes affect locations ahead of the transient's actual position.

Care has to be taken in assessing the effect of linear interpolation when comparing MoC predictions with site- or laboratory-recorded pressure surge data. Generally the MoC simulations underestimate the frictional resistance contribution to attenuation (Vardy 1976; Swaffield 2010), so that the rounding errors associated with linear interpolation with TFAC >1 may at first sight improve the simulation/test data comparison. This is erroneous, and as a rule the lowest possible value of TFAC should be used. It is appreciated that this may lead to difficulties if pipes within the network have widely varying wave speeds, and utilising a moderate TFAC is the only means available to ensure that the time step remains constant for all system pipes and satisfies the Courant Criterion for each.

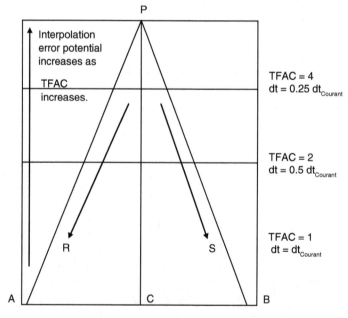

Figure 3.10 Potential for linear interpolation errors increases as the time step decreases relative to its initial value as the flow velocities within the network vary with appliance discharge. Note the necessity to retain a constant time step across the whole network to facilitate junction calculations implies that the time step may be artificially reduced in a drain with little or no flow, implying a high risk of interpolation errors.

In the case of free surface transient flow simulation particular problems are encountered in the drawdown and backwater profiles upstream of a free outfall in a subcritical flow or a junction or obstruction imposition of a local subcritical flow regime within an otherwise supercritical flow condition. Swaffield (1982) identified an interpolation problem characterised by the collapse of a backwater profile before an upstream wave reached the channel exit. Bridge (1984) introduced a holding mechanism that retained the back-water profile that will be discussed later.

The use of a polynomial fit through the three nodes upstream of the channel exit was proposed in order to improve the interpolation where the water surface has a predominant curvature – for example upstream of a free outfall in a shallow gradient, high frictional resistance channel, Figure 3.11. The four coefficients required for the cubic equation are determined from the known values of flow depth and velocity at adjacent nodes at time t together with estimates of the gradient of depth and velocity with respect to distance at these two adjacent nodes determined from the general expression for the gradually varied free surface profile, equation 2.31,

$$\frac{dx}{dh} = \frac{1 - QT/gA^3}{S_0 - Q^2/C^2mA^2} = \frac{1 - Fr^2}{S_0 - S_f} \tag{3.49}$$

where the numerator is the expression for Critical depth and the denomina-tor the expression for Normal depth. The alternative form involves inclusion

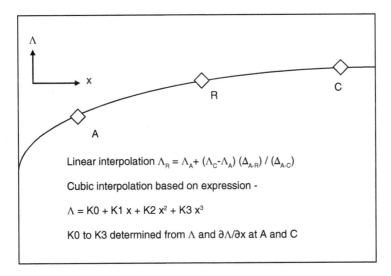

Linear interpolation $\Lambda_R = \Lambda_A + (\Lambda_C - \Lambda_A)(\Delta_{A-R})/(\Delta_{A-C})$

Cubic interpolation based on expression -

$\Lambda = K0 + K1\,x + K2\,x^2 + K3\,x^3$

K0 to K3 determined from Λ and $\partial\Lambda/\partial x$ at A and C

Figure 3.11 Cubic interpolation technique (Murray 1982). Comparison of linear and cubic based interpolation techniques.

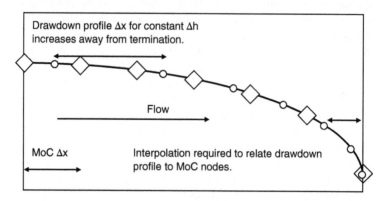

Figure 3.12 Interpolation to yield MoC node depths introduces initial errors as distance increments do not match.

of the flow Froude Number, Fr, and the channel friction slope S_f, defined previously in equations 2.3 and 2.28.

This then allows interpolated values at intermediate points to be determined based as before on the time step utilised. It should be noted here that equation 3.49 allows the determination of dx for constant values of dh; thus the value of dx will increase rapidly as the integration moves upstream along the drawdown profile, Figure 3.11. Thus points at which depth are known accurately will not conform to a pre-set Δx grid, which will introduce a further interpolation error in setting up the initial steady flow base condition for any simulation. The inverse of equation 3.49 allows the local surface gradient to be determined upstream of the Critical flow depth location.

Generally the St Venant equations of continuity and momentum are cast in terms of u and h as the variables describing the flow conditions. However there is a case for considering the application of the so-called 'conservation form' of the St Venant equations where u and h are replaced by Q and A in the expectation that the combined variable Q will display a more linear variation. In steady flow this is naturally true as the flowrate is constant; however the application is worthy of consideration as in any unsteady flow the local velocity and depth will be subject to greater individual rates of change than the combined flowrate variable.

The characteristic equations for free surface partially filled pipe flow with no lateral inflow may be written, from equation 3.25 and Table 3.5 as

$$\frac{du}{dt} \pm \frac{g}{c}\frac{dh}{dt} + g(S - S_0) = 0 \qquad (3.50)$$

provided that $\dfrac{dx}{dt} = u \pm c.$

The conservation form of the characteristic equation pair, in terms of Q and A, therefore become expressed as

$$\frac{d(Q/A)}{dt} \pm \frac{g}{c}\frac{dA}{Tdt} + g(S - S_0) = 0$$

as $dA = Tdh$ (Figure 3.3), so that

$$\frac{1}{A}\frac{dQ}{dt} - \frac{Q}{A^2}\frac{dA}{dt} \pm \frac{g}{cT}\frac{dA}{dt} + g(S - S_0) = 0$$

and as $c = \sqrt{gA/T}$ the expression becomes

$$\frac{dQ}{dt} - \frac{Q}{A}\frac{dA}{dt} \pm c\frac{dA}{dt} + gA(S - S_0) = 0$$

provided that

$$\frac{dx}{dt} = \frac{Q}{A} \pm c$$

Under steady flow conditions the flowrate at each node upstream of a free outfall in a subcritical flow will have a constant value, Q, and therefore no interpolation error would be expected. Under these conditions the depth at discharge will be the flow Critical depth, given by the expression

$$1 - \frac{Q^2 T_c}{gA_c^3} = 0$$

so that given T(h) and A(h) for any cross-sectional shape there will be an unique solution for discharge flow and depth.

However, in a free surface flow simulation, periods of steady flow will be interspersed with unsteady, and therefore the prediction technique has to be the same for both. This requires interpolation to determine both Q and A at an upstream location, and while the Q prediction may be improved by use of the conservation form of the St Venant equations, the prediction for A will be subject to the same potential errors as previously. Thus the simultaneous solution of the flow characteristic and the Critical depth boundary may return a set of Q, h values that satisfy the Critical depth boundary but still display an error compared to the known steady flow condition.

Vardy and Pan (1998) reviewed interpolation techniques in MoC applications and concluded that linear interpolation was 'safe' in as much as the interpolated values would always lie between the adjacent node values. The cubic formulation was considered problematic due to difficulties in obtaining accurate values of the gradient terms at each adjacent node, *'the potential to improve greatly on linear interpolation is accompanied by a corresponding potential to be much worse'* (Vardy and Pan 1998). Unlike linear interpolation, the cubic approach is not inherently 'safe'.

The MoC solution of the St Venant equations of continuity and momentum is an industry standard approach that has applications across the whole field of pressure transient propagation. Its popularity is undoubtedly based on the already discussed perception of the technique segregating the internal calculations from the design of boundary conditions. The MoC development of boundary conditions for solution with the appropriate characteristic is a mechanism that encourages innovative boundary condition development and the multi-role application of simulations with small additions for changes in boundary condition. However interpolation is an explicit component of the MoC and therefore the accuracy and success of any simulation is bound to be highly dependent on the accuracy of the interpolation technique.

However, while it may be accepted that MoC is perhaps not the most accurate of simulation techniques and has significant shortcomings in some applications, it retains its position as a method of first choice due to its compartmentalisation of the modelling necessary and the advantages offered by the freedom to design innovative boundary conditions.

In the development of the DRAINET simulation (Swaffield 1982; Bridge 1984; Swaffield and Galowin 1992), interpolation issues were encountered in setting up the initial base conditions and during the time period between the inception of unsteady flow at the entry to any network branch and the arrival of the surge waves at hydraulic jumps upstream of junctions and obstructions to flow. In particular it was found that the initial steady subcritical backwater profiles upstream of junctions and obstructions tended to collapse during this period, an effect caused by the imposition of variable time steps increasing the effective TFAC number and hence increasing the interpolation errors (Figure 3.10). Equation 3.35 defined the Courant Criterion time step as

$$\Delta t = \frac{\Delta x}{(u + c)_{max}}$$

where the maximum value of u+c is the maximum found at any node in the network at that time. The value of the time step will therefore change at each new time step. It has already been stressed that the time step has to remain the same for all branches in a network whether the branch carries an appliance discharge or merely the initial setup trickle flow.

Bridge (1984) introduced a correction factor approach which has been found to be a pragmatic solution to the interpolation issue for free surface unsteady flows where the rate of change of the flow parameters is mild.

Referring to Figure 3.6d, under steady flow conditions the values of u, h and c at A, B and C remain constant with time as do the interpolation values at R, S and S′ necessary to give a set of values at P identical to those at C a time step earlier. Bridge introduced a correction factor into the characteristic equations for free surface flow, equations 3.51 and 3.52 derived from Table 3.6, where S stands for S′ too,

$$u_P = u_R - \frac{g}{c_R}[h_P - h_R] - g[S_R - S_0]\Delta t = K1 - K2h_P \tag{3.51}$$

$$u_P = u_S + \frac{g}{c_S}[h_P - h_S] - g[S_S - S_0]\Delta t = K3 + K4h_P \tag{3.52}$$

Correction factors C_1 and C_2 which may be defined as

$$C_1 = \left[-u_C + u_R - \frac{g}{c_R}[h_C - h_R] \right] / g[S_R - S_0]\Delta t \tag{3.53}$$

$$C_2 = \left[-u_C + u_S + \frac{g}{c_S}[h_C - h_S] \right] / g[S_S - S_0]\Delta t \tag{3.54}$$

may be proposed that effectively hold the backwater profile during steady flow periods. C_1 and C_2 may be included in the K1 and K3 terms so that the characteristic equations become

$$u_P = u_R - \frac{g}{c_R}[h_P - h_R] - C_1 g[S_R - S_0]\Delta t = K1 - K2h_P \tag{3.55}$$

$$u_P = u_S + \frac{g}{c_S}[h_P - h_S] - C_2 g[S_S - S_0]\Delta t = K3 + K4h_P \tag{3.56}$$

These correction factors only apply within the subcritical backwater profiles upstream of junctions or obstructions, in the subcritical drawdown profile upstream of a free outfall or in the Critical depth to normal supercritical flow development profile downstream of a junction or branch entry. At all other nodes the value of both C_1 and C_2 may be set to 1.0. The C_1 and C_2 constants may be seen to effectively change the frictional representation in the subcritical profile zones. During the initial steady flow period the subcritical flow becomes frictionless and the initial values are held. Once unsteady flow conditions are imposed the friction loss is under-represented with the consequent effect that the incident wave suffers reduced attenuation and therefore retains a higher propagation velocity c. However the proportion of the network affected will be small as the backwater profiles upstream of junctions are inevitably short. The normal flow regime within the drainage network would be supercritical, nodes within that regime being allocated values of unity for both C_1 and C_2 constants.

As shown in Figure 3.13 it is necessary to set up the network with an initial trickle flow, QBASE, at time zero. This is set within DRAINET at between 0.01 and 0.1 litres per second. In order to avoid this flow accumulating downstream the simulation accepts that each branch will have an initial flow of QBASE, i.e. the flow continuity error at each junction is accepted. Thus the initial time step based on equation 3.48 will be 'large' even for internodal reaches of typically 0.25 to 0.5 m. As the discharge to the network changes, so the value of Δt will change, reducing as the flowrate increases. The stabilising constants both C_1 and C_2 include the initial time step Δt determined at time zero. Therefore for C_1 and C_2 to be effective in retaining a gradually varied flow profile prior to the arrival of a wave from upstream or as a backflow from a junction, the value of Δt must be kept stable for this period. The only practical technique to achieve this result is to set the initial Δt below the Courant Criterion level, effectively by introducing a

TFAC value greater than unity, writing the time step as dt $= \dfrac{dx}{TFAC(u+c)_{max}}$;

values in the range 1–3 have been used satisfactorily. If the applied discharges result in a time step less than the pre-set value then the simulation would automatically utilise the new time step. The interpolation errors initially

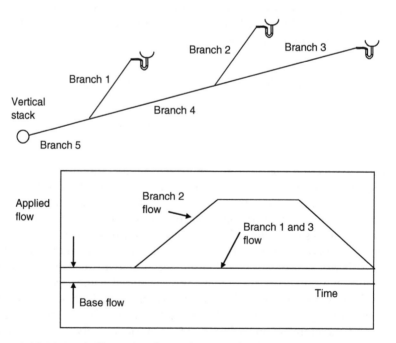

Figure 3.13 Network illustrating the need to monitor branch flows and adjust the network time step to satisfy the Courant Criterion.

expected from the use of a TFAC value >1 obviously disappear once the time step falls to a level below its pre-set value, as the time step will then be at the Courant Criterion value. When the discharge abates and the time step rises to become greater than the pre-set value, then the simulation holds the time step at its initial value. This ensures that the correction factors again aid the re-establishment of a steady flow condition once all the network discharges have dissipated and the network flow is again merely the pre-set trickle flow.

Maxwell Standing (1986) introduced a comprehensive improvement by using the Everett and Newton-Gregory interpolation techniques to accurately model the backwater surface curvature, as will be discussed in the later treatment of rainwater gutter flows. Maxwell Standing (1986) reviewed the quadratic and Lagrangian interpolation techniques previously utilised to attempt to improve the interpolation issue (Lister 1960; Mozayeng and Song 1969; Jolly and Yevjevich 1971). Jolly and Yevjevich found that Lagrangian techniques improved the interpolation relative to a linear scheme but that there was no advantage in utilising more sophisticated schemes than third order linked to an iterative determination of the positions of R and S. Maxwell Standing determined that interpolation schemes that involve central differences are the most efficient for a given computational effort and concluded that the practical interpolation formulae are those attributed to Everett and Bessel, with the Everett formulae being preferred for these applications based on level of third-order differences.

Everett's formulae only involve even-order central differences, the scheme being expressed as

$$P(x) = E_0 f_0 + E_2 \partial^2 f_0 + E_4 \partial^4 f_0 + \dots\dots\dots\dots\dots\dots$$
$$+ F_0 f_1 + F_2 \partial^2 f_1 + F_4 \partial^4 f_1 + \dots\dots\dots$$

$$E_0 = 1 - u \qquad F_0 = u$$

$$E_2 = \frac{-u(1-u)(2-u)}{3!} \qquad F_2 = \frac{u(u^2-1)}{3!}$$

$$E_2 = \frac{-(-1-u)u(1-u)(2-u)(3-u)}{5!} \qquad F_2 = \frac{u(u^2-1)(u^2-4)}{5!} \qquad (3.57)$$

where $u = \dfrac{x - x_0}{\Delta x}, 0 < u < 1$ and $\partial^i f_n$ is the ith central difference at $x = x_n$.

Close to branch entry and exit where central differences cannot be obtained, Maxwell Standing utilised the Newton-Gregory forward and backward difference formulae:

$$P(x) = P(x_0 + hu) = f_0 + u\Delta f_0 + \frac{1}{2}\left[u(u-1)\Delta^2 f_0\right] + \frac{1}{3!}\left[u(u-1)(u-2)\Delta^3 f_0\right]$$

$$+\dots\dots\dots\dots + \frac{1}{n!}\left[u(u-1)\dots\dots(n-n+1)\Delta^n f_0\right]$$

and

$$P(x) = P(x_0 + hu) = f_0 + u\nabla f_0 + \frac{1}{2}\left[u(u+1)\nabla^2 f_0\right] + \frac{1}{3!}\left[u(u+1)(u+2)\nabla^3 f_0\right]$$

$$+................+\frac{1}{n!}\left[u(u+1).....(u+n-1)\nabla^n f_0\right] \tag{3.58}$$

Thus three interpolation schemes are available to provide the inter-polated characteristic base conditions within the free surface simulation DRAINET. These will be referred to as LIN, C1C2 and NGE in the following comparisons.

Figure 3.14 illustrates the attenuation of an appliance discharge along the length of branch drain, representing the influence of drain slope and surface roughness. It is noticeable that within 10 m the peak flow depth predicted has fallen to roughly a third of its maximum value. The attenuation will be shown later to be dependent upon both drain and discharge characteristics, including drain diameter, surface roughness and slope and the initial base flow and the rate of change of the applied appliance discharge. These rela-tionships will be returned to later.

The choice of time step has already been discussed. In order to retain a constant time step across a network it may be necessary to reduce the time step below the initial Courant Criterion value. Figure 3.15 illustrates the effect of introducing a TFAC value of 3 into the time step expression,

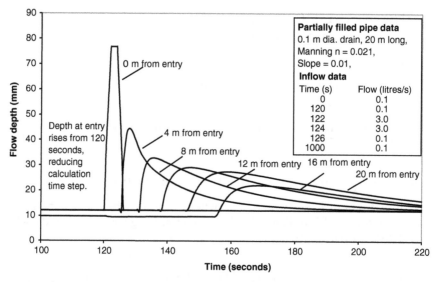

Figure 3.14 Attenuation of an appliance discharge along a 20 m length of 100 mm diameter branch set at a 0.01 slope, illustrating the decrease in wave height.

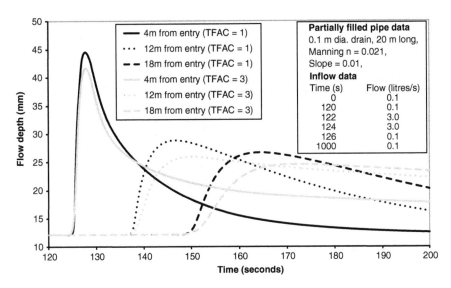

Figure 3.15 Influence of time step choice on the predicted wave attenuation along the 20 m branch drain.

equation 3.48. The peak flow depth predicted is little changed; however there is a marked 'spreading' of the wavefront following the passage of its leading edge – as predicted in the earlier discussion of possible interpolation rounding errors. While this may not be of major significance for the prediction of peak depths, it may have an influence on solid transport predictions based on the relationship between solid velocity and the local predicted water velocity.

Figure 3.16 illustrates the success of each interpolation scheme in holding the exit control depth ahead of the arrival of the appliance discharge wave. As discussed, the C1C2 (Bridge 1984) correction factor holds the exit depth at the Critical value; however once the appliance discharge commences at 120 seconds the time step decreases sharply as the determining flow velocity and wave speed rise, equation 3.48. As expected increasing the TFAC factor leads to further rounding errors. The Newton-Gregory and Everett interpolation equations are the most efficient in retaining the exit depth; however there is further interpolation rounding in this scheme too as the imposed time step decreases with rising inflow hydrograph. On the basis of Figure 3.16 it is clear that the NGE scheme is the more efficient, both in terms of holding the initial flow depth profile and in reducing the fall off once the appliance discharge commences.

It was mentioned earlier that the LIN and C1C2 schemes implied that the wave front was predicted to travel faster than the actual wave speed due to the interpolation effect. Figure 3.17 illustrates this effect. The NGE scheme predicts a rise in flow depth at the branch exit some 3 seconds later than that for either the C1C2 or LIN schemes. It may be seen that the NGE scheme does introduce

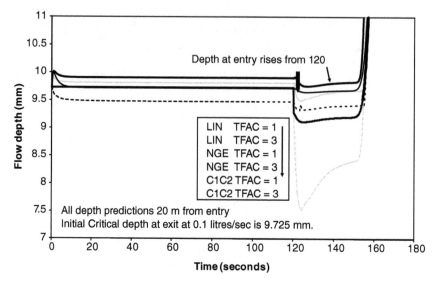

Figure 3.16 Influence of interpolation scheme on the retention of the branch drain exit Critical depth value ahead of the arrival of the appliance discharge wave.

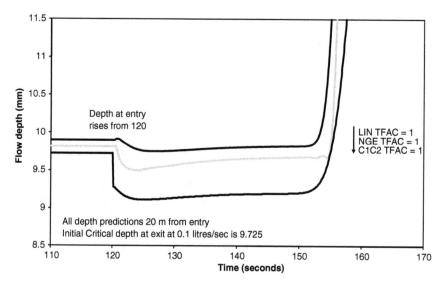

Figure 3.17 Influence of the chosen interpolation scheme on the rate of change of depth following the arrival of the incident appliance discharge wave.

small oscillations at the arrival time of the incident wave; however the depth increase is sharper than that predicted by either of the other techniques.

The applied time step, equation 3.48, must always be equal to or less than the Courant Criterion value. In some cases it will be necessary to reduce the time step by use of a TFAC > 1 in order to ensure a constant time step across a network where there are wide divergences between the flows carried by each branch. Figure 3.18 illustrates the change in time step that follows the appliance discharge. It will be seen that as soon as the appliance discharge commences at 120 seconds the time step falls in line with equation 3.48. As the discharge abates the time step recovers, again following equation 3.48.

The choice of interpolation scheme also affects the subsequent attenuation of the wave as mentioned above. Figure 3.19 illustrates the peak predicted depth 4 m from entry for both the LIN and NGE schemes. It will be seen that the peak depth is little affected, but that there is a less severe attenuation as the TFAC value is increased. Clearly maintaining the highest possible TFAC value consistent with maintaining a constant time step for the network is the preferred choice.

The influence of the choice of internodal reach length should also be considered. The influence of internodal reach length on the interpolation issue is small as the errors in interpolation are a function of the geometry, e.g. the process, and therefore are not overly affected by increasing the number of nodes per m. This confirmed earlier observations (Swaffield 1981a) during the initial development of the DRAINET model.

Table 3.7 presents a comparison of the minimum time step and maximum flow depth and flowrate predicted 8 m from entry in a 100 mm diameter drain set at a slope of 0.01 and having a Manning n value of 0.21 subjected to the appliance discharge detailed in Figure 3.15, a discharge with a peak inflow of 3 litres/second. The comparisons of peak predicted depth and flowrate confirm that the choice of the number of nodes per m is of

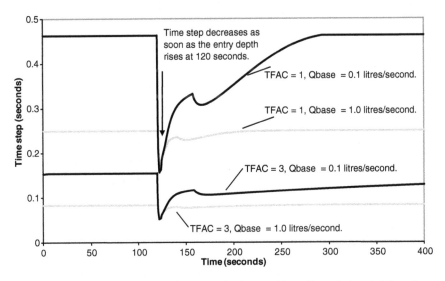

Figure 3.18 Time step variation based on both initial base flow, 0.1 or 1.0 litres/second, and TFAC value, 1 or 3.

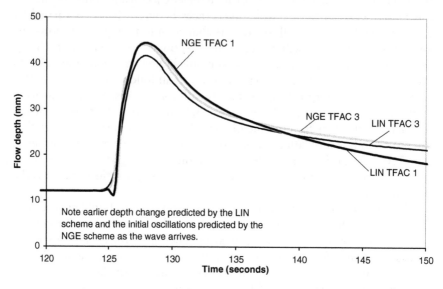

Figure 3.19 Flow depth 4 m from entry as predicted by both the LIN and NGE interpolation schemes.

Table 3.7 Comparative depth and flowrate predictions as the number of nodes per m is increased from 1 to 16. Note this implies a 256 increase in the number of calculations as the number of nodes also affects the time step utilised

Number of nodes per m	1	2	4	8	16
Initial Δt	0.9979	0.4989	0.2494	0.1247	0.0624
Minimum Δt	0.6611	0.3327	0.1654	0.0828	0.0417
Max. depth 8 m from entry, mm	51.84	50.74	50.27	50.06	49.82
Max. flowrate 8 m from entry, litres/second	2.02	1.934	1.901	1.889	1.883

little consequence at the calculation densities proposed. Two nodes per m provides a reasonable representation of the network while retaining discrimination in terms of defining the inflow profiles per time step.

3.3 Alternate simulation schemes

The preceding text has detailed the application of the Method of Characteristics to address the simulation of unsteady free surface flows. This presentation drew upon experience of the advantages of this technique in dealing with pressure transient applications in a whole range of full bore conditions, ranging from hydroelectric station analysis to low-amplitude air pressure transients in building drainage systems (Swaffield and Boldy 1993; Swaffield 2010). The difficulty in application to free surface flows has been discussed, namely the issues surrounding initial water surface prediction, the necessity to introduce an initial trickle flow and interpolation and rounding errors due to the curvature of the surface profiles close to imposed entry or exit conditions. These difficulties have been addressed and shown to be manageable.

However the majority of large-scale free surface channel simulations tend to utilise finite difference techniques to simulate unsteady flow, albeit in much larger channels than the partially filled drainage pipes and open rainwater gutters addressed here. There is an extensive literature available; however it is predominantly concerned with larger-scale civil engineering flood routing and open channel hydraulics (Abbottt and Minns 1998; Pender *et al.* 2005). The advantages of the MoC approach in simulating unsteady flow under both subcritical and supercritical conditions have been recognised, emphasising that most open channel application would be subcritical – not the case in the treatment of building drainage networks where the flow conditions are predominately supercritical with the exception of backwater profiles imposed by junctions or defects.

Despite the application of finite difference techniques to general open channel free surface flows, no application to the analysis of internal building drainage systems is available. Kamata, Matsuo and Tsukagoshi (1979) suggested the use of the Lax-Wendorff finite difference technique to model free surface flows in a drainage branch with the upstream inflow as a function of time being used to drive the simulation. Although wave attenuation was demonstrated, there does not appear to have been any further development of the simulation, and the finite difference treatment was merely used to confirm the wave attenuation demonstrated by Swaffield and Marriott (1978). More recently the simulation of the transport of finite waste solids in the sewer network has become of interest (Butler and Davies 2004); however in the sewer network the flow is acknowledged as quasi-steady so that the flow simulation is sufficiently accurately provided by the HydroWorks commercial package developed to aid sewer flow prediction. A similar approach was followed by Lauchlan, Griggs and Escarameia (2004). The DRAINET model utilising the MoC solution of the St Venant equations therefore remains the only available simulation technique, based on the MoC discussion presented in this chapter and summarised in Swaffield and McDougall (1996).

The application of CFD to free surface unsteady flow at the scale of building drainage systems has been initiated by the company Geberit in 2007. The transport of waste solids was represented in a simple 5 m branch drain flowing discharge from a single w.c. The run time for the process was stated to be several days, and the fluid mechanisms were represented by standard CFD packages. It is clear that as computing capability continues to develop at an exponential rate such simulations with three-dimensional capability will become attractive and helpful, particularly if the complex flow conditions in the vicinity of junctions, including the possibility of hydraulic jump translation and backflow are modelled. However that is some way in the future. The DRAINET simulation offering a more simplistic one-dimensional solution of the St Venant equations would take several seconds of laptop computing to complete the same task. Interestingly, when DRAINET was first developed by Swaffield and Galowin at NBS Washington in 1980, runs times in hours were experienced and mainframe computers were required – an indication of the simulation infrastructure. The issue of the suitability of CFD package approach will be returned to later; however it is important to recognise the suitability of any chosen technique in terms of the accuracy expected of the output and the accuracy of the input data.

In building drainage system operation the base data is often poorly known as, for example, the actual appliance discharge to be modelled may vary depending upon random factors such as the mass of faecal material to be flushed. Thus it is perhaps more appropriate at this stage to consider the ability of simpler models to quickly map out the general range of output that might be expected rather than commit to a single simulation run taking an inordinate time to deliver a single prediction. CFD packages will develop in the future and will become available. Their development is to be encouraged; however at the time of writing it would appear that the MoC solution, drawing on 40 years of recognition as an industry standard simulation technique offers the better modelling option.

In contrast to internal building drainage systems, a number of finite difference–based models have been developed to simulate conditions within rainwater drainage systems. The main driver behind the development of such models has primarily been the increased use of siphonic rainwater drainage systems, where the increased complexity and consequences of failure necessitate more detailed design techniques than traditional empirically based approaches. Good examples of siphonic system simulation models include the SIPHONET and ROOFNET models developed at Heriot-Watt University and the SYFON model developed at the University of South Australia (Beecham *et al.* 2008).

The remainder of this section presents a brief overview of finite difference approaches, before going on to detail the application of one such approach to rainwater drainage systems.

3.3.1 Finite difference methodology

Finite difference solution schemes to the governing St Venant equations replace the partial differential terms with finite difference approximations to these terms. Whilst it is possible to develop a large number of different operators, normally three basic types are employed, namely forward, backward and central differencing. With respect to Figure 3.20, a finite difference approximation to the term $\frac{\partial z}{\partial x}$ may be expressed as:

Forward differencing: $\dfrac{\partial z}{\partial x} = \dfrac{z_{j+1}^n - z_j^n}{\Delta x}$ (3.59)

Backward differencing: $\dfrac{\partial z}{\partial x} = \dfrac{z_j^n - z_{j-1}^n}{\Delta x}$ (3.60)

Central differencing: $\dfrac{\partial z}{\partial x} = \dfrac{z_{j+1}^n - z_{j-1}^n}{2\Delta x}$ (3.61)

It is important to remember that these approximations are based on 'truncated' Taylor's series expansions, where higher-order terms are ignored. For example, forward and backward differencing are said to be first-order accurate as they neglect second- and higher-order terms, resulting in a truncation error of:

$$\frac{d^2 z}{dx} \frac{\Delta x^2}{2!} \pm \frac{d^3 z}{dx} \frac{\Delta x^3}{3!} \pm \frac{d^4 z}{dx} \frac{\Delta x^4}{4!} \pm \dots$$ (3.62)

Similarly, central differencing is second-order accurate, as it neglects third- and higher-order terms. Approximating the differential equation terms therefore introduces an error whose value is dependent upon the internodal grid size chosen.

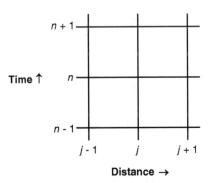

Figure 3.20 Space-time grid as a basis for a finite difference scheme, where n represents time progression and j represents the nodal number, increasing in the initial flow direction.

It is important to note that, irrespective of the type of finite difference technique used, the conditions at internal and external boundaries are always calculated using the principles of characteristics.

3.3.2 Explicit/implicit schemes

Finite difference solution techniques can be classified as either explicit or implicit. Essentially the difference lies in the information used in the finite difference operators. Explicit schemes use only known information (i.e. that at the current time step) and can thus be solved directly, whilst implicit schemes use a combination of both known and unknown information (i.e. that at the current and next time step) and must therefore utilise some form of iterative solution technique. This is illustrated by equations 3.63 and 3.64, which show both an explicit and an implicit finite difference approximation to the term $\dfrac{\partial z}{\partial x}$:

$$\text{Explicit:} \quad \frac{\partial z}{\partial x} \approx \frac{\left(z_{j+1}^{n} - z_{j}^{n}\right)}{\Delta x} \tag{3.63}$$

$$\text{Implicit:} \quad \frac{\partial z}{\partial x} \approx \theta \frac{\left(z_{j+1}^{n+1} - z_{j}^{n+1}\right)}{\Delta x} + (1-\theta)\frac{\left(z_{j+1}^{n} - z_{j}^{n}\right)}{\Delta x} \tag{3.64}$$

Both explicit and implicit methods have advantages and disadvantages. Neither can be treated as straightforward, as both require a degree of understanding of the likely consequences of poor time and internodal distance choices. Explicit methods have the advantage of being relatively easy to program and apply, but the need to comply to the Courant condition often necessitates short time steps, which inevitably leads to longer computer run times. In contrast, whilst implicit methods are generally more stable and can thus use longer time steps, they are intrinsically more complex as all nodal point calculations must be undertaken simultaneously. The resulting numerical models of necessity use matrix methods of solution and are cumbersome and are in detail beyond the scope of this text.

3.3.3 Conservation form of the St Venant equations

Stability is a continuing concern in the modelling of unsteady flow conditions. It has been shown that in the explicit formulations it is necessary to tightly control time and internodal step size to reduce the possibility of unstable solutions. Criteria such as the Courant Criterion have been introduced. Instability may also be caused by poor choice of variables to define flow conditions – for example in any unsteady flow the local velocity and depth will be subject to greater individual rates of change than the combined flow rate variable. Hence many studies have employed the 'conservation form' of the St Venant equations, effectively replacing velocity and depth with volumetric flowrate and area as the two variables to be determined against time and distance. The generalised form of the continuity and momentum equations applicable to gradually varied unsteady flow in an open channel are given by equations 3.2 and 3.9. Neglecting lateral inflow, these equations may be developed into the conservation form for unsteady flow in open channels thus:

$$\frac{\partial A}{\partial t} + \frac{\partial Q}{\partial x} = 0 \tag{3.65}$$

$$\frac{\partial Q}{\partial t} + \frac{\partial}{\partial x}\left[\frac{Q^2}{A} + gI\right] - \left[\begin{array}{c} 0 \\ gA(S_o - S_f) \end{array}\right] = 0 \tag{3.66}$$

where I_1 is the first moment of the area A about the free surface.

3.3.4 MacCormack explicit finite difference solution

The MacCormack solution is a variation of the classical Lax-Wendorf (1960), which is one of the most frequently found techniques to employ a finite difference representation of the conservation form of the St Venant equations; this technique is superior to the MoC as it has the capability to pass a surge or hydraulic jump. It relies on the hyperbolic nature of the governing equations, which leads to spontaneous discontinuities that have real physical meanings, e.g. hydraulic jumps.

The MacCormack technique is a non-centered, two-step finite difference scheme which is second-order accurate in time and space. Starting from the initial time level, when steady state theory is used to calculate conditions at regular points throughout the system, the solution at the next time level is computed in a two-step *predictor–corrector* process. This technique is illustrated schematically in Figure 3.21 and may be described by first expressing the governing equations in vector form, thus:

$$\frac{\partial U}{\partial t} + \frac{\partial F}{\partial x} = S \tag{3.67}$$

where: $U = \left[\begin{array}{c} A \\ Q \end{array}\right];\quad F = \left[\begin{array}{c} Q \\ \dfrac{Q^2}{A} + gI_1 \end{array}\right];\quad S = \left[\begin{array}{c} 0 \\ gA(S_o - S_f) \end{array}\right]$

Application of the MacCormack technique now enables the solution to proceed as follows:

1   From known flow variables at time n, the values of U_i^n, F_i^n and S_i^n can be calculated for all nodes.

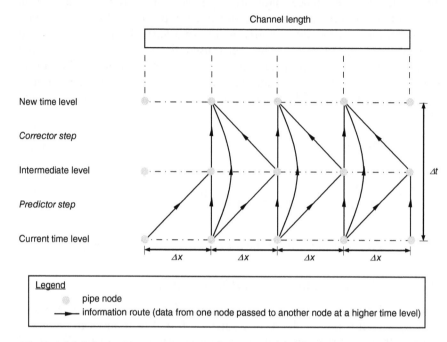

Figure 3.21 Schematic representation of one version of the MacCormack method.

2   Predictor values $\left(U_i^p \right)$ can then be calculated for each node from

$$U_i^p = U_i^n - \frac{\Delta t}{\Delta x}\left[(1-\varepsilon)F_{i+1}^n - (1-2\varepsilon)F_i^n - \varepsilon F_{i-1}^n\right] + \Delta t S_i^n \qquad (3.68)$$

3   The predictor values F_i^p and S_i^p can then be calculated for each node from the predictor values U_i^p.

4   Using the predictor values (again with known U_i^n values at time n), the U_i^{n+1} values and hence flow variables at time n+1 can then be calculated from

$$U_i^{n+1} = 0.5\left(U_i^n + U_i^p\right) - \frac{\Delta t}{2\Delta x}\left[\varepsilon F_{i+1}^p + (1-2\varepsilon)F_i^p + (\varepsilon-1)F_{i-1}^p\right]$$
$$+ \frac{\Delta t}{2}S_i^p \qquad (3.69)$$

In the preceding expressions, the ε term may be set equal to either zero or one, and is used to vary the direction of differencing; e.g. if ε is set to 1 forward differencing is used during the predictor stage, whilst backward differencing is used during the corrector step (this is the approach shown in

Figure 3.21). Several authors have discussed the choice of difference expression, and current opinion favours either the alternate use of forward or backward expressions in the predictor and corrector steps of the solution or aligning the difference expression to the direction of motion of any wave moving through the system (Fennema and Chaudhry 1989).

Although the approach outlined here utilises the free surface St Venant equations, it can be used to model full bore flow conditions by introducing a Preissmann slot in the crown of the pipe (Fennema and Chaudhry 1989) (see Figure 3.22). Ideally the width of the Preissmann slot should be such that the wave speed in the composite, and fictitious, pipe/slot section equals that which would occur in the actual pipe under the same conditions. The width of the slot (w) has been shown (Yen 1986) to be given by an expression that includes pipe diameter and the wave speed in the full bore flow simulated (c), thus:

$$w = \frac{\pi D^2 g}{4c^2} \tag{3.70}$$

Typically this would yield extremely narrow slot widths, which would result in numerical instabilities. Consequently it is normally necessary to adopt a larger slot width, generally of the order of 1% to 5% of the pipe diameter. Although this approach introduces a small degree of computational error, it is a reasonable compromise for numerical stability.

As the MacCormack method is explicit, it must satisfy the Courant Criterion at each node at each time step. In addition, the second-order nature of the MacCormack method means that it generates spurious oscillations in the vicinity of hydraulic jumps: oscillations which have no real meaning and are purely a numerical anomaly. There are a number of methods of alleviating this problem, including adding artificial viscosity to the solution scheme and incorporating Total Variation Diminishing (TVD) schemes (Toro and Titerev 2005).

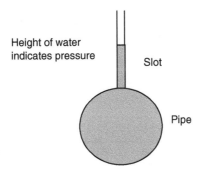

Figure 3.22 Illustration of the Preismann slot to allow simulation of pressurised pipe flow.

3.4 Boundary conditions applicable to internal building drainage systems

It has been shown that the development of both MoC and finite difference solutions for unsteady flow simulation may be regarded as having two specific and separate calculation elements. Internal nodes are solved either by the simultaneous solution of the available C^+ and C^- characteristics, in the case of MoC solutions, or by solution of finite difference approximations to the St Venant equations in finite difference solutions. At system boundaries, be they internal or external, conditions are determined using *boundary conditions*; irrespective of the solution technique applied, these boundary conditions will make use of the principle of characteristics. Any unsteady flow simulation is ultimately only as good as the representation of the system boundary conditions, and it is the rate of change of conditions at the channel entry and exit that governs the wave propagation.

The following section will concentrate on the definition and development of boundary conditions applicable to internal building drainage systems, including entry and exit boundary conditions as well as internal moving boundary conditions. In order to establish the technique for the identification of suitable boundary equations it is probably best to start by considering entry conditions.

3.4.1 Entry boundary conditions

Entry conditions depend upon the flow regime expected along the branch drain and the discharge category. Some appliances, such as baths, showers, sinks, may be assumed to have a relatively tranquil entry flow condition continuing for a considerable time (Figure 3.23) and may be simulated by reference to either a Critical or Normal depth entry condition. The discharge

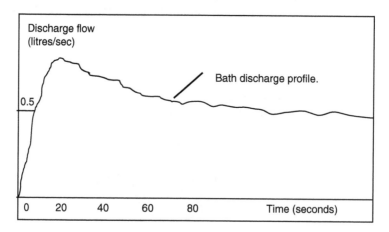

Figure 3.23 Typical quasi-steady bath discharge profile to a horizontal branch drain via an appliance trap seal.

from w.c. appliances has sufficient kinetic energy to impose a short time, shallower, rapid flow condition at drain entry that may be simulated by reference to the flow kinetic energy.

Similarly the entry flow to a drain from a vertical stack that may have accumulated the discharge from several upper floor networks has to take into account both the modification of the floor discharge profiles during the vertical annular flow experienced in the stack and the high kinetic energy content of the flow when it reaches the vertical stack to horizontal stackbase connection.

Figure 3.24 illustrates an appliance trap seal to branch drain connection. If the downstream flow condition is supercritical then no C^- characteristic is available at drain entry and the flow condition is imposed, either via the Critical depth, equation 2.21, or the Normal depth, equation 2.22, solved with the known appliance discharge expressed as an array representing an the discharge profile $Q = f(t)$ curve. Simulations have shown that assuming an imposed Normal depth is adequate as the depth development profile from Critical to Normal depth will be short in most cases.

If the downstream flow regime is expected to be subcritical then the available C^- characteristic is solved with either the Critical or Normal depth relationships. In each case the output will be an entry depth at each time step that responds to the appliance discharge profile.

The discharge of a w.c. results in a short time duration unsteady inflow to the drain with a higher energy content so that the flow depth at drain entry is noticeably lower than either the Critical or Normal depths for the applied discharge (Figure 3.25). It was therefore necessary to characterise the w.c. discharge flow through laboratory testing. The w.c. discharge is idealised as

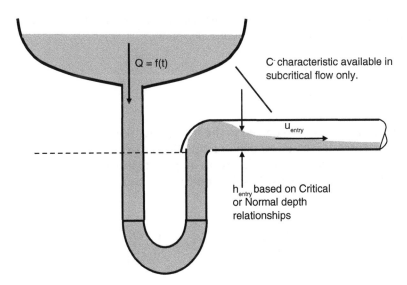

Figure 3.24 Entry to a branch drain from an appliance trap seal.

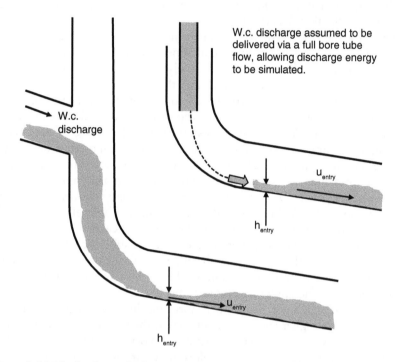

Figure 3.25 Idealised w.c. discharge to a branch drain to represent the energy
content of the inflow.

a full bore flow delivery where the discharge is known as a $Q = f(t)$ function,
and the flow kinetic energy therefore becomes

$$KE_{w.c.} = \frac{1}{2g}\frac{Q_t^2}{A_{tube}^2} \tag{3.71}$$

This energy may be equated to the free surface flow specific energy, S.E.,
at drain entry. It is assumed that in most design cases the w.c. discharge will
enter a branch that would, due to its slope, support supercritical flow so
that the solution is imposed as the branch entry condition. Thus the energy
balance may be stated as

$$SE_{entry} = h_{entry} + \frac{u_{entry}^2}{2g} = K(KE_{w.c.}) = K\frac{1}{2g}\frac{Q_t^2}{A_{tube}^2} \tag{3.72}$$

where K is an energy loss coefficient normally taken to have the value 0.5.

Equation 3.72 may be solved at each time step as the applied flowrate is known from $Q = f(t)$; the form of a typical w.c. discharge is illustrated in Figure 3.25. The idealised delivery tube diameter was set at 40 mm; however this value may require reconsideration in the future as w.c. design develops. The introduction of low flush volume w.c.'s would not necessarily affect the ideal tube diameter as the peak flow achieved by new designs still has to deliver good appliance performance.

3.4.2 Exit boundary conditions

As shown in Figures 3.6 (a to d) there will always be an available C^+ characteristic available at exit linking upstream conditions known at point R at time t to exit conditions at point P at time $t+\Delta t$. In the case of a free outfall from a free surface supercritical flow this equation may be solved with the also available C^- characteristic linking conditions at S' at time t to exit conditions at point P at time $t+\Delta t$. Under subcritical free surface conditions only the C^+ characteristic is available to be solved with the expression for Critical depth at the free outfall, hence the available equations are

$$u_{N+1} = K3 + K4h_{N+1} = 0 \tag{3.73}$$

and for Critical depth, as $1 - \dfrac{Q^2 T_c}{gA_c^3} = 0$, it follows that the Critical exit velocity may be expressed as

$$u_{N+1} = \frac{Q_{N+1}}{A_{N+1}} = \sqrt{\frac{gA_{N+1}}{T_{N+1}}} \tag{3.74}$$

which of course is also the wave speed as the Froude number is unity at Critical depth. Solution of equations 3.73 and 3.74 may be achieved by the bisection technique as both A and T are known functions of h.

3.4.3 Junction boundary conditions

If the channel terminates in a junction of two or more branches then a boundary condition must be found to describe the junction depth. At a junction of n branches, n–1 being deemed upstream and terminating at the junction and the other branch naturally having the junction as its entry condition, there are n unknown values of each of the variables u and h to be determined; hence six equations are required. If branches 1 and 2 terminate at the junction then, if the enhanced depth at the junction due to the confluence of flows has imposed a local subcritical flow regime, it follows that C^+ characteristics exist for branches 1 and 2, linking in each case the known conditions at the upstream point R at time t to unknown conditions at the junction at time t+Δt. The downstream entry into branch 3 will be governed by the Critical depth equation for the combined flow entering branch 3. The termination depths in branches 1 and 2 will depend upon an empirical expression defining the junction flow depth in terms of the combined approach flows (Swaffield and Galowin 1992), and the final equation will be provided by flow continuity through the junction.

Thus the required equations may be expressed as

$$u_{1,N+1} = K3 + K4h_{1,N+1} = 0 \, , \, u_{2,N+1} = K3 + K4h_{2,N+1} = 0$$

$$\sum_{i=1}^{3} Q = \sum_{i=1}^{2} u_{i,N+1}A_{i,N+1} + u_{3,1}A_{3,1} = 0$$

$$h_{1,N+1} = h_J, \, h_{2,N+1} = h_J, \, u_{3,1} = \sqrt{\frac{gA_J}{T_J}}.$$

Swaffield and Galowin (1992) presented empirical junction depth vs. combined approach flowrate expressions for both 90° and 135° level invert junctions (Figure 3.26),

$$h_{Junction} = 0.039Q^{0.5714} \text{ for a } 90° \text{ level invert junction and} \qquad (3.75)$$

$$h_{Junction} = 0.035Q^{0.5714} \text{ for a } 135° \text{ level invert junction.} \qquad (3.76)$$

The relationship between combined flowrate and downstream Critical depth (Figure 3.20), independent of pipe slope and roughness, may be approximated by

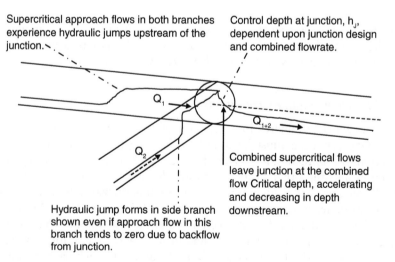

Figure 3.26 Depth at a level invert junction.

$$Q = 3.06D^{0.567}h_c^{1.95} \qquad (3.77)$$

so that the junction depth vs. combined flowrate expressions become

$$90° \text{ level invert junction } h_{\text{Junction}} = 0.074D^{0.324}h_c^{1.114} \qquad (3.78)$$

$$135° \text{ level invert junction } h_{\text{Junction}} = 0.066D^{0.324}h_c^{1.114} \qquad (3.79)$$

expressions in terms of the flow Critical depth, which is independent of pipe slope and roughness and readily determined.

The flow cross section and surface width are again known as functions of depth. Normally these relationships are drawn from the geometry of the circle; however non-circular section drains may be treated in the same way provided the cross-sectional geometry is defined (Cummings, McDougall and Swaffield 2007).

Swaffield and Galowin (1992) also applied the concept of the Critical depth as a determinant of junction depth for top entry junctions (Figure 3.27), proposing that a relationship of the form

$$\frac{h_{\text{Junction}} - h_{mc}}{h_{mc}} = K\left\{\frac{h_{dc} - h_{mc}}{h_{mc}}\right\}^n \qquad (3.80)$$

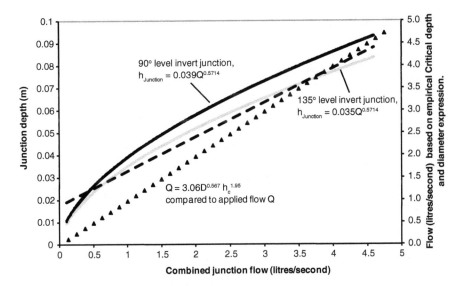

Figure 3.27 Depth relationships determined experimentally for two level invert junctions.

would be a suitable boundary equation, where suffix mc indicates the level invert upstream approach flow Critical depth and dc indicates the combined flow downstream Critical depth. Laboratory test rig measurements indicated that for a 90° top entry junction the values of K and n were typically 2.15 and 0.56.

In the case of a top entry junction the exit condition for the discharging upper branch is provided by the solution of the available C^+ and C^- characteristics as the branch will in all probability be under conditions of supercritical flow. The remaining four equations are provided by the main drain C^+ characteristic, the continuity of flow equation, the downstream combined flow Critical depth expression and the junction depth expression, in terms of flow Critical depths.

3.4.4 Obstruction boundary conditions

In the case of a branch terminating in an obstruction then the same boundary condition approach may be followed by considering the obstruction as a two-branch in-line junction where the relationship between approach flow and the obstruction depth replaces the junction depth expression. Downstream branch entry will again be at the appropriate flow Critical depth. Empirical data for a 20 mm high obstruction in a 100 mm diameter drain (Swaffield, McDougall and Campbell 1999) suggested that the flow depth over the obstruction was given by an expression of the form:

$$h_{obs} = z + ((1.71S_0 - 0.0026)z - 63.0S_0 + 1.43)h_c$$
$$+ (10.9S_0 - 0.87)z + 450.0S_0 + 3.36 \qquad (3.81)$$

where all dimensions are in mm.

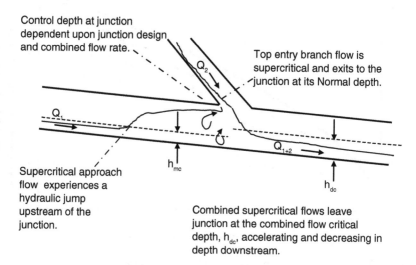

Figure 3.28 Depth at a top entry junction.

While the obstruction effect was dependent upon an empirical measurement of a particular obstruction, the case of a defective slope may be handled by reference to gradually varied flow theory to determine the flow depths and the location of any upstream hydraulic jump.

Figure 3.29 illustrates a typical slope defect where the slope varies from steep to mild as shown and then reverts to steep. The flow would therefore be expected to transition from supercritical via a hydraulic jump, whose upstream location would depend on the junction depth, $h_{junction}$, to subcritical and then back to supercritical via a Critical depth control, h_c, downstream of the defective slope section. The backwater profile linking these control depths would depend upon the drain parameters of slope, diameter and roughness as well as the throughflow Q, determined via the general gradually varied flow integral, equation 3.82:

$$\frac{dh}{dx} = \frac{S_0 - Q^2 / C^2 m A^2}{1 - QT / gA^3} = \frac{S_0 - S_f}{1 - Fr^2} \qquad (3.82)$$

The initial depth profile is thus determined by integrating the gradually varied flow depth equation 3.82 back from the Critical depth control to yield the junction depth. As the simulation proceeds it is necessary to undertake this profile determination within the solution for depth and velocity at the junction based on the predicted approach flow. This is clearly an iterative solution involving a throughflow based on the available C^+ characteristic. Depth and velocity values in the defective pipe section are determined in the usual way from the available C^+ C^- characteristics.

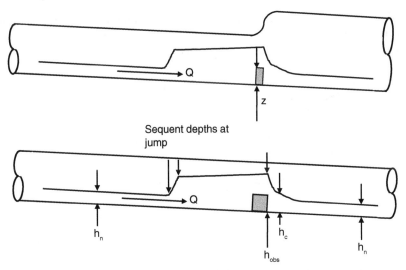

Figure 3.29 Flow depths upstream and downstream of an obstruction or a badly made in-line pipe junction.

3.4.5 Vertical stack boundary conditions

The discharge of a vertical stack to a collection drain may be approached in a similar fashion to the energy boundary condition for a w.c. discharge. Again the flow will be highly unsteady, possibly of short duration. Figure 3.30 illustrates the accumulation of discharge flows down a vertical stack following appliance discharges to the networks on each upper floor. It will be appreciated that this combined discharge profile will include both short-duration unsteady flow periods, corresponding to w.c. discharge, and relatively quasi-steady interludes resulting from the discharge of baths, showers etc.

Figure 3.31 illustrates the annular downflow in the vertical stack and its transition to a free surface flow in the horizontal drain. The annular flow may form a water curtain across the entry to the horizontal drain with consequences for the propagation of air pressure transients in the drainage and vent system (Swaffield 2010). The transition from annular to free surface horizontal flow will naturally involve an energy transfer so that the effective energy available at drain entry is reduced. It will be assumed that within the annular flow the kinetic energy along any streamline is based on the annular downflow terminal velocity, equation 2.36, namely $V_t = K(\frac{Q_w}{D})^{0.4}$ where the value of K is dependent upon stack roughness.

Assume that the energy available in the terminal annular flow may be expressed as

$$e = \frac{V_t^2}{2g}(\sin^2 0.5\theta) \qquad (3.83)$$

Equation 3.83 implies that at $\theta = 0$ the vertical annular flow loses its entire kinetic energy and at $\theta = \pi$ the streamline energy is preserved. The annular flow specific energy may then be expressed as

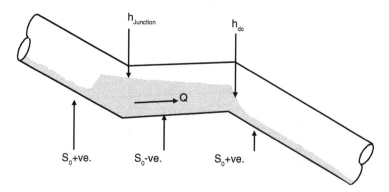

Figure 3.30 Flow depths upstream and downstream of a slope defect in a horizontal branch. Note that convention dictates that a downwards pipe slope is taken as positive.

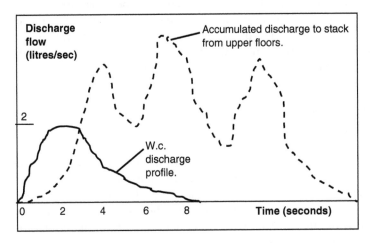

Figure 3.31 Discharge profiles accumulate at the base of a vertical stack to establish a combined discharge to the downstream drain.

$$\int_0^{2\pi} e\left(\frac{D}{2}-t\right) t d\theta = KE_B \frac{\pi}{4}\left(D^2 - (D-2t)^2\right) \tag{3.84}$$

where KE_B is the bulk annular kinetic energy. Substituting for e, equation 3.84, allows E_B to be determined as

$$KE_B \frac{\pi}{4}\left(D^2 - (D-2t)^2\right) = 2\int_0^{\pi}\left(\frac{D}{2}-t\right) t \frac{V_t^2}{2g}\left(\sin^2\frac{\theta}{2}\right) d\theta$$

$$= \left(\frac{D}{2}-t\right)\frac{t}{g} V_t^2 \int_0^{\pi}\left(\sin^2\frac{\theta}{2}\right) d\theta$$

$$= \left(\frac{D}{2}-t\right)\frac{t}{g} V_t^2 \left(\frac{\pi}{2}\right)$$

and if t $\ll$ D as would be the case in a typical vertical stack it follows that:

$$KE_B = \left(\frac{D/2-t}{D-t}\right)\frac{V_t^2}{2g} = K_{stackbase}\frac{V_t^2}{2g} \Rightarrow 0.5\frac{V_t^2}{2g} \tag{3.85}$$

(Note that the design condition for a vertical stack limits the annular thickness to between 1/9th and 1/16th of the stack diameter; the value of the loss coefficient in these case would be 7/16 and 7/15, so 0.5 is acceptable as a general application).

The entry depth at the stackbase may then be determined as for a w.c. discharge by equating the annular flow kinetic energy, including the loss coefficient, to the entry flow specific energy. Again it is assumed that in the vast majority of simulations the downstream drain flow regime will be supercritical and the entry depth and velocity may be determined from equation 3.86.

$$SE_{entry} = h_{entry} + \frac{Q_{drain}^2 / A_{drain}}{2g} = K_{stackbase} \frac{V_t^2}{2g} \qquad (3.86)$$

Figure 3.31 illustrated the accumulation of flow down a vertical stack. Each floor discharges its appliances to the stack so that the annular downflow becomes the summation of a series of $Q = f(t)$ profiles, each representing the discharge from a particular floor to the stack. This accumulated flow then provides the overall Q–$f(t)$ input data that drives the simulation of the horizontal drain free surface flow.

The difficulty is that the individual floor discharges will be altered in terms of the shape of each $Q = f(t)$ profile as flows with differing magnitudes will assume different terminal velocities in the stack. In addition the confluence of a floor discharge profile with the downward flow upper floors will again modify the stack flow profile.

It is therefore necessary to introduce some modelling of the vertical stack annular flow.

Figure 3.32 introduces a methodology designed to modify the discharge profile to the vertical stack to represent its development as the terminal velocity, dependent upon stack parameters and the applied flowrate, either to increase the steepness of the rising profile or spread the decreasing sections of the overall discharge flow profile. The DRAINET simulation utilises this technique to link floors vertically and therefore allow a full building simulation, including the wave attenuation in the drain downstream of the stackbase. Clearly the time increments used in the simulation would be considerably shorter than those indicated in Figure 3.32 so that the accumulated profile at stackbase tends to be a continuous curve.

The incident profile at each floor is represented by a finite number of time steps, the flow rate varying within each, in Figure 3.32 increments AB to EF as shown. For each flow at points A to F the flow terminal velocity and annular thickness may be determined from the annular terminal velocity, equation 2.36, $V_t = K(\frac{Q_w}{D})^{0.4}$ where $K = 0.632/n^{0.6}$ and $t = Q/(\pi DV_t)$.

The arrival time of each of the identified profile points A to F at the next floor may then be determined at the time taken for a fluid particle to fall a distance Δz to be

$$\Delta t = V_t \Delta z \qquad (3.87)$$

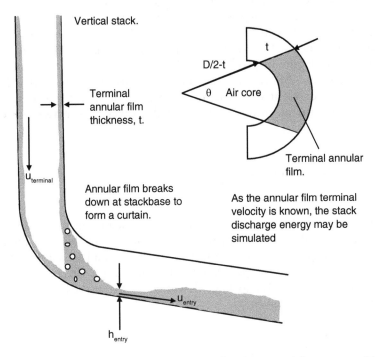

Figure 3.32 Transition from annular to free surface horizontal flow at a stackbase.

A new profile may then be constructed to represent the flow profile arriving at the next floor below at times from A' to F', If the interfloor distance is insufficient for all terminal conditions to develop then some approximation to the acceleration curve may be included. Wise (1973) reported distances to terminal velocity from 5 to 7 m, and therefore it would be prudent to include a factor to represent this development length – linear would appear to be sufficiently accurate when the other unknowns are considered; hence let the assumed annular velocity $V_{annular}$ be approximated by

$V_{annular} = ZF\ V_t$ where $ZF = \Delta z\ /\ 6$ if $\Delta z < 6$ and $ZF = 1$ if $\Delta z \geq 6$ so that the general expression for the translation of a point on the inflow profile becomes

$$t_i' = t_i + \Delta z\ /\ (ZF\ V_t) \qquad (3.88)$$

The translation of the inflow profile illustrated in Figure 3.32 relies on the following 'rules':

1 Constant inflow sections, e.g. AB, CD or EF. Here the arrival time duration remains constant so that $\Delta t(AB) = \Delta t(A'B')$ etc. The arrival time of

the leading edge of the profile at A is given by equation 3.88, while the arrival flowrate arriving between A'B' remains $Q_{AB} = Q_{A'B'}$

2   Increasing inflow at the upper floor, e.g. section BC. In this case the profile point C will travel faster than point B so that $\Delta t(BC) < \Delta t(B'C')$. The arrival time of profile point C at the lower floor is again given by equation 3.88. As there has been a reduction in the arrival Δt for the flow represented by the area beneath line BC, it is necessary to determine the mean flow arriving at the lower floor between B'C' as follows:

$$Q_{B'C'} = \frac{1}{2}(Q_B + Q_C)\frac{T_B - T_C}{T_{B'} - T_{C'}} \tag{3.89}$$

This steepens the wave front as the arrival time increment is reduced.

3   Reducing inflow at the upper floor, section DE. In this case the trailing profile point E travels slower than point D and so the flow profile spreads as $\Delta t(DE) > \Delta t(D'E')$, the arrival time of profile point C is given by equation 3.88. The average flowrate arriving at the lower floor during ΔDE is again given by

$$Q_{D'E'} = \frac{1}{2}(Q_D + Q_E)\frac{T_D - T_E}{T_{D'} - T_{E'}} \tag{3.90}$$

This spreads the wave front as the arrival time increment is increased.

Figure 3.32 illustrates the modified profile A'F' arriving at the lower floor, and this profile may then be combined with the branch to stack inflow profile at that floor represented by the profile OZ. This process then repeats at each lower floor until the final modified profile arrives at the stackbase to provide the local $Q = f(t)$ that will be used in conjunction with equation 3.86 to provide the drain entry condition.

The summation mechanism was developed for use with the DRAINET model (Swaffield Bridge and Galowin 1982); however it was hoped to improve the simulation by undertaking a study of the annular stack flow to develop an MoC simulation that could be incorporated into DRAINET. Thancanamootoo (1991) developed a form of the St Venant equations relevant to the annular flow regime:

$$\frac{\partial V}{\partial \tau} + V\frac{\partial V}{\partial z} - g + f\frac{V^2}{2t} + \frac{P}{\rho t}\frac{\partial t}{\partial z} = 0 \text{ for momentum} \tag{3.91}$$

$$V\frac{\partial t}{\partial z} + \frac{\partial t}{\partial \tau} + t\frac{\partial V}{\partial z} = 0 \text{ for continuity} \tag{3.92}$$

$$\frac{dV}{d\tau} + \lambda \frac{dt}{d\tau} - \left(g - f \frac{V^2}{2t} \right) = 0$$

where

$$\lambda = \pm \frac{1}{t} \sqrt{\frac{P}{\rho}}$$

and

$$\frac{dz}{d\tau} = V \pm \sqrt{\frac{P}{\rho}}$$

Note in these equations t is annular thickness, τ is time, P is absolute air pressure in the stack, effectively atmospheric, and ρ is water density.

As with other MoC applications it was necessary to develop boundary conditions at entry and exit from any interfloor stack section. The model was found to be unsatisfactory at entry to the stack due to the rapid and non-linear reduction in annular thickness during the development of the annular film. Swaffield and Thancanamootoo (1991) improved the model by introducing conformal transformation to allow an increased node density close to the stack entry which improved the mass balance to within acceptable limits, <1%, however it was clear that this model would not be sufficiently robust to be included in the DRAINET simulation, which therefore retains the summation technique discussed above and illustrated by Figure 3.32.

3.4.6 Hydraulic jump boundary conditions

Under initial steady flow conditions the position of a hydraulic jump upstream of a junction or defective drain condition has been shown to be determined by reference to the gradually varied flow expressions, in particular the integral expression equation 2.31.

$$dx_{h_1-h_2} = \int_{h_1}^{h_2} \frac{1 - Q^2 T / gA^3}{S_0 - Q^2 / C^2 mA^2} dh = \int_{h_1}^{h_2} \frac{1 - Fr^2}{S_0 - S_f} dh \qquad (3.93)$$

Under unsteady flow conditions the jump is free to move upstream from the junction or obstruction control as the approach flow increases or back towards the control as the approach flow abates. In addition, an initial jump located in one of the other joining branches not experiencing any change in the upstream applied flow conditions will move in response to backflow from the junction. It is therefore necessary to include the jumps in all joining branches as moving boundary conditions within the DRAINET simulation.

At first sight this is a straightforward case of identifying the relevant equations. Across the jump there are five unknowns to be determined at any time during the simulation, namely the depth and velocity of the flow upstream and downstream of the jump and the velocity of the jump itself as it responds to the surrounding flow conditions.

The five equations available may be identified as

1   The C^+ characteristic upstream of the jump in the Δz reach AC, Figure 3.33,

$$u_P = u_R - \frac{g}{c_R}[h_P - h_R] - g[S_R - S_0]\Delta t = K1 - K2h_P \qquad (3.94)$$

2   The C^- characteristic upstream of the jump in the Δz reach AC, Figure 3.33,

$$u_P = u_S + \frac{g}{c_{S'}}[h_P - h_{S'}] - g[S_{S'} - S_0]\Delta t = K3 + K4h_P \qquad (3.95)$$

3   The C^- characteristic downstream of the jump in the Δz reach CB, Figure 3.33,

$$u_P = u_S + \frac{g}{c_S}[h_P - h_S] - g[S_S - S_0]\Delta t = K3 + K4h_P \qquad (3.96)$$

4   The continuity of flow equation written across the jump location,

$$(u_1 - u_w)A_1 = (u_2 - u_w)A_2 \qquad (3.97)$$

5   The momentum equation for the jump,

$$\rho g A_1 \bar{h}_1 - \rho g A_2 \bar{h}_2 = \rho A_2 (u_2 - u_w)^2 - \rho A_1 (u_1 - u_w)^2 \qquad (3.98)$$

where it is assumed that the jump length tends to zero, eliminating any shear force effect, and that the channel slope may be ignored and that centroid depth is again given by equation 2.33.

$$\bar{h} = X0 + h - \frac{D}{2} = 0.666 \frac{D}{2} \frac{\left(3\sin\frac{\theta}{2} - \sin 3\frac{\theta}{2}\right)}{4\left(\frac{\theta}{2} - \frac{1}{2}\sin\theta\right)} + h - \frac{D}{2} \qquad (3.99)$$

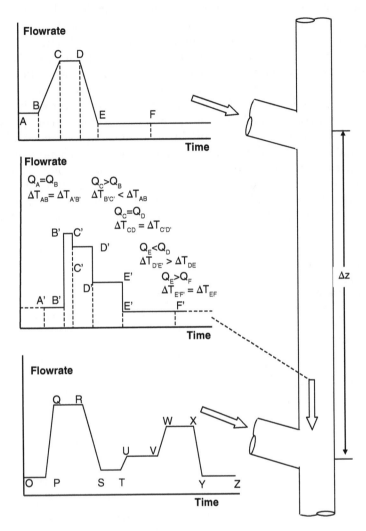

Figure 3.33 Flow accumulation in a vertical stack serving a number of upper floors.

Equations 3.70 to 3.73 are the standard characteristic equations met previously; however as they apply to Δx sections on either side of the jump as it moves through the regular characteristic grid, it will be appreciated that depth and velocity values at the nodes A, B, C are interpolated from the nearest regular grid nodes, introducing potential rounding errors as already discussed. The equations 3.74 and 3.75 apply to rapidly varying flow and are valid across the jump as it moves through the grid. Solution of these five equations allows the velocity of the jump to be determined as well as the flow and depth on either side. The passage of the jump through the grid may be tracked by introducing a pseudo characteristic with a slope $1/u_w$ as illustrated for both jumps upstream of a junction in Figure 3.34. The DRAINET simulation includes this solution mechanism, and the movement of the jump may be predicted, including its response to back flow from a junction.

The treatment of the hydraulic jump introduces the concept of a moving boundary condition whose defining equations refer to the surrounding flow conditions. A second example of a moving boundary would be a discrete solid being transported through the network by the attenuating appliance discharge flows. Solid transport will be returned to in detail in a later chapter; however it is sufficient at this stage to emphasise that solid interaction will become important and that a suitable model may also be developed through the use of a velocity decrement model that relates solid properties to the velocity differential between the solid and the predicted surrounding free surface flows.

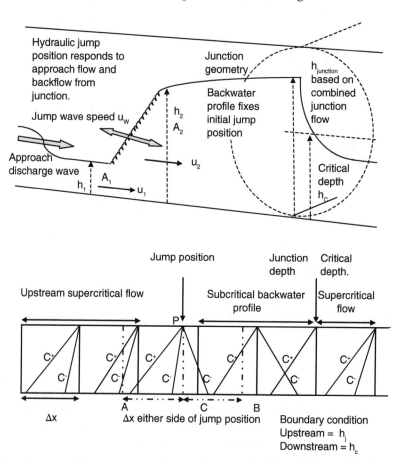

Figure 3.34 Boundary conditions necessary to simulate the movement of a hydraulic jump within branches terminating at a junction or a drain defect.

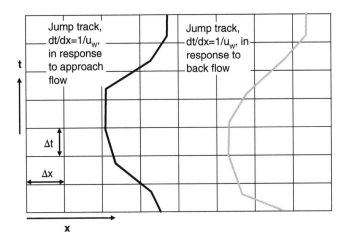

Figure 3.35 Hydraulic jump response to an increasing approach flow and subsequent junction backflow, followed by a return to initial conditions as appliance discharge abates.

3.5 Boundary conditions applicable to rainwater drainage systems

Just as with internal building drainage systems, boundary conditions are required to determine the conditions at external and internal boundaries within rainwater drainage systems. A number of the boundary conditions for rainwater drainage systems are equally applicable to those already developed for internal building drainage systems, including: subcritical and supercritical system exit, level entry junction (for horizontal pipework in siphonic systems), vertical stack flow (for downpipe flows) and hydraulic jump (for horizontal pipework in siphonic systems when using MoC solution technique). However, additional boundary conditions are required to represent elements specific to rainwater drainage systems. These boundary conditions will be developed in Chapter 6, following an overview of the characteristics of the various different types of rainwater drainage system available.

3.6 Concluding remarks

This chapter has introduced the Method of Characteristics as a solution technique applicable to the St Venant equations of continuity and momentum defining unsteady free surface flows in building drainage networks. The development of the technique has drawn heavily on its use as the industry standard methodology to simulate pressure surge across a whole range of engineering fluid mechanics applications. The central importance of the Courant Criterion in

Table 3.8 Summary of the boundary equations developed for the MoC simulation of unsteady free surface flows in building drainage systems and rainwater gutters

Boundary condition	Solution methodology
Unsteady inflow at drain entry 1 Subcritical flow	1 Solve the available C– characteristic with the inflow profile by introducing either the Q-dependent Critical or Normal depth.
2 Supercritical flow, entry from an appliance or a vertical stack	2 Impose flow depth and velocity at entry by reference to appliance discharge profile and equate flow kinetic energy to the free surface specific energy, having made allowances for local energy 'losses'.
3 Entry of a long time base appliance discharge, e.g. sink, bath, shower	3 Impose Critical or Normal depth for supercritical flows, solve the Critical or Normal depth equation with the available C– characteristic for subcritical flows.
4 Entry flow downstream of a junction or obstruction into downstream supercritical flow	4 Impose the Critical depth based on the junction throughflow.
Dead ended channel	Impose flow velocity = zero and solve with available C+ characteristic.
Gutter discharge to a restricted outfall	As gutter flow is predominantly subcritical solve available C+ characteristic with a local depth vs. Q expression representing the degree of outfall restriction.
Lateral inflow to gutter or land drain	Include lateral inflow term in the St Venant equations and hence in both the C+ and C– characteristics.
Free outfall 1 Supercritical flow 2 Subcritical flow	1 Solve available C+ and C– characteristics. 2 Solve available C+ characteristic with the Critical depth equation.

Level invert junctions	Solve available C+ characteristic in each joining branch with the empirical junction depth vs. combined throughflow expression.
Top entry junctions	1 Lower branch – solve as for level invert junction but including top entry flowrate. 2 Upper discharging branch will be steep, so solve as for a supercritical free outfall.
Moving hydraulic jump upstream of a junction of obstruction	Solve available C+ and C– characteristics for approach flow and C– characteristic for subcritical flow immediately downstream of the jump with continuity of flow and momentum equations across the jump.
Obstruction to flow	Solve the available C+ characteristic with an empirical relationship linking the depth over the obstruction to the approach flowrate.
Slope defect 1 Upstream 2 Downstream	1 Solve the available C+ characteristic at entry to the defect with the local flow depth determined from the backwater profile through the defect. 2 Impose the Critical depth appropriate to the defect throughflow.

ensuring stability through control of the calculation time step has been emphasised and detailed.

The advantages of the MoC solution have been discussed, in particular the relatively simple concept that all the required calculations fall into two broad categories, namely simultaneous solution of the available characteristics at each internal drain length node and boundary calculations involving some defined boundary equation solved with, if possible, one available characteristic. It has been shown that the flow regime, subcritical or supercritical, determines the form of the boundary condition solutions as well as introducing the concept of moving boundary conditions within the drain length to represent the translation of hydraulic jump position in response to either an unsteady approach flow or backflow from a downstream junction. Each of these cases is capable of solution.

The use of the MoC solution for free surface flows introduces a range of issues, including the necessity to provide an initial base flow over which the later unsteady wave will run, as well as the necessity to determine initial steady flow depth profiles in the vicinity of branch drain entry and exit conditions. In addition the issues of rounding errors and interpolation problems were discussed, and several interpolation schemes were presented and shown to progressively alleviate these concerns. Linear interpolation for characteristic base conditions was compared to more complex Newton-Gregory and Everett interpolation expressions and shown to be suitable.

In addition to the MoC, this chapter has also introduced alternative simulation schemes to the governing St Venant equations, concentrating primarily on the MacCormack finite difference technique. The benefits of such schemes and associated stability constraints are detailed. Importantly, it is shown that, even when using a non-MoC based simulation approach, the principles of characteristics form the basis for boundary conditions.

Thus Chapter 3 has established the simulation methodologies and has drawn upon the basic steady flow treatment in Chapter 2 to allow predictions of wave attenuation, system response and solid transport in later chapters. Chapters 4 to 6 will apply these simulation methodologies to the prediction of a range of system responses to various operational profiles and will utilise fully the base theory developed in Chapters 2 and 3.

Note

1 Adhémar Jean Claude Barré de Saint-Venant (b. August 23, 1797 Villiers-en-Bière, Seine-et-Marne; d. January 1886) was a mathematician who developed the one-dimensional unsteady open channel flow shallow water equations or Saint-Venant equations that are a fundamental set of equations in modern hydraulic engineering. He taught mathematics at the École des Ponts et Chaussées, where he succeeded Coriolis. In 1868 he was elected to the Académie des Sciences and continued research work for a further 18 years.

4 Simulation of free surface unsteady flow in building drainage networks

The flow regime within building drainage networks is both dominated by, and arguably defined by, the attenuation of appliance discharges as they propagate through the network. Attenuation is apparent in the observed reduction in peak flowrate and depth as the free surface waves generated by an appliance discharge travel through the network with the effect that, at some distance downstream, the flow appears quasi-steady, characterised by reduced flow depth, reduced flowrate and an extended flow duration. Once this zone is reached the simulation of wave action, for example by such techniques as the already discussed Method of Characteristics, is no longer necessary, and more simplistic quasi-steady flow models may be used.

Wave attenuation is a function of the differential wave speed and flow velocity along both the leading and trailing edges of the wavefront, as illustrated in Figure 4.1. An observer would recognise that the observed peak wave depth, or flowrate, reduces downstream leading to the often remarked upon condition that downstream drains, examined for example at an open manhole, display very little flow as a result of an upstream appliance discharge. However the duration of flow is observed to increase markedly. Figure 4.1 illustrates this effect while Figure 4.2 defines the mechanism in terms of differential wave speed.

For a triangular discharge profile, chosen here to emphasise the effect, the leading edge steepens while the trailing edge flattens, effectively lengthening the overall wave disturbance in terms of both length and the time taken to pass any point along the drain. However for the trailing edge to flatten, the peak depth or flowrate must decrease as shown. As the wave is inelastic the only mechanism to allow this to occur is a redistribution of the fluid within the wave – thus the leading edge peak is reduced while the wavelength increases.

In the case of a discharge profile displaying a plateau, Figure 4.2 presents a trapezoidal profile as an example. It follows by the same mechanism that, while the leading edge steepens, the peak leading edge maximum depth or flowrate does not decrease until the flattening of the training edge 'erodes' the wave profile plateau, as shown by reference to points AA' and BB'.

The wave profile attenuates in the flow direction, both h and Q peak values decrease while flow redistribution within the wave extends the overall length of the disturbance.

The leading edge may steepen while the trailing edge always flattens due to the relative values of the wave speed, c, and the flow velocity, u.

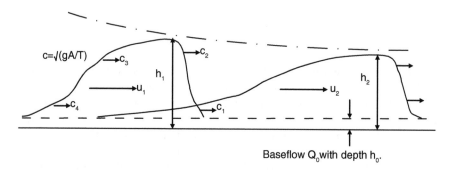

Figure 4.1 Mechanism of wave attenuation in a partially filled channel.

Therefore the modelling of wave attenuation becomes essential to understanding and predicting the flow regime within the building drainage network. The inherent importance of wave attenuation was recognised by Hunter (1940), whose comment that the steady state equations only apply to drainage flow for very short periods of time remains central to the attenuation discussion. However Hunter also implicitly recognised that no then available mathematical model existed to capitalise on the reduced downstream flow loading. Wyly (1964) also recognised empirically the effect of wave attenuation by publishing maximum flowrates for particular drainage conditions of slope and diameter in excess of the norms predicted by the then current codes, the enhanced drain flow capacity being dependent upon the flow duration – a clear application of attenuation theory. Burberry (1978) measured wave attenuation experimentally by recording the time taken for a 9 litre w.c. flush to pass points along a 100 m diameter drain at a range of slopes. His results led to his comments that for drains in excess of 10 m major increases in acceptable flowrate were possible and that there was a need to develop a 'computing solution' to enable attenuation to be incorporated into drainage design. These considerations and opinions from leading members of the research community over 30 years supported the development of the DRAINET MoC simulation (Swaffield and Galowin 1992), initially at NBS and then both Brunel and Heriot-Watt Universities in the UK.

The attenuation-driven reduction in peak flowrate and depth is also an essential component in the mechanisms determining solid transport within drainage networks. It will be shown later that predictions of wave attenuation

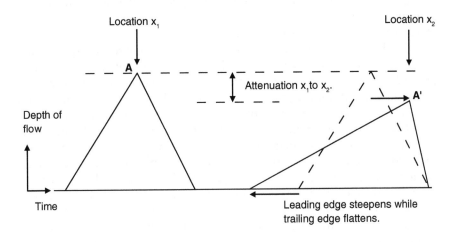

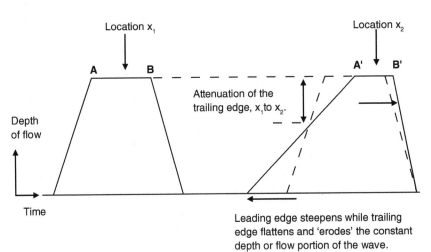

Figure 4.2 The attenuation mechanism is dependent upon the form of the applied wave.

via the MoC will allow the velocity and energy content of an appliance discharge wave to be simulated and hence allow both the velocity of discrete solids within the discharge wave and the minimum flowrate of depth necessary to initiate the continuing motion of a deposited solid to be determined.

It is necessary initially to identify the factors determining wave attenuation, and this may be achieved by considering the mechanisms already described and the effect of three sets of system parameters upon these mechanisms.

Wave attenuation will depend upon the partially filled pipe characteristics, namely diameter, roughness and slope. In addition Figure 4.2 has indicated that the 'shape' of the discharge profile is itself a determining parameter.

Figure 4.2 also indicates, although not explicitly, that the attenuation is determined by the baseflow over which the discharge wave propagates as this affects the driving differences in wave speed.

It will be shown later, during a consideration of non-circular branch drain wave attenuation, that the cross-sectional shape also affects attenuation, however this is again implicit in Figure 4.2 as the value of wave speed depends at any flow depth on the ratio of cross-sectional area to wetted perimeter. Generally in this chapter the conduit cross section will be assumed circular.

While wave attenuation will suggest that the capacity of a drainage network may be underestimated if no account of wave attenuation is included, the presence of junctions within the network will provide a counterbalancing argument. The transition from supercritical to subcritical flow through the mechanism of the hydraulic jump will result in flow depth immediately upstream of junctions increasing rapidly in the flow direction. It is therefore necessary to give particular consideration to the design of junctions. It will be appreciated that, as the junction depth will have a singular value regardless of how many branches come together at the junction, backflow will occur into any branch that is not carrying an approach flow. This effect will be demonstrated, and experimental validation provided, as will the possibility of reverse flow carrying solids back into the joining branch drains where long-term deposition may occur.

In order to allow a whole building simulation to be undertaken Chapter 3 has outlined a mechanism to allow the summation of upper floor discharges to a vertical stack to provide a stackbase discharge profile to drive unsteady flow and wave attenuation in the sewer connection. DRAINET simulations will be presented to illustrate the efficacy of this technique and the resulting wave attenuation on the sewer connection, demonstrations backed up by experimental validation of the model.

4.1 Dependence of wave attenuation on drain and flow parameters

The wave attenuation in a partially filled circular–cross section length of drain will be shown to be dependent upon the drain diameter, roughness and slope, the form of the discharge profile and the baseflow present in the drain ahead of the appliance discharge. The attenuation predictions will be provided by application of the DRAINET simulation utilising the boundary conditions already developed in Chapter 3. The DRAINET simulation (McDougall and Swaffield 2000) incorporates a graphical user interface that links system representation to the boundary condition selection, as illustrated in Figure 4.3.

The DRAINET simulation was used to consider the attenuation of discharges to the network having profiles as illustrated in Figure 4.4 – namely peak and trapezoidal profiles chosen to aid a definition of the dependence of attenuation on discharge type. Table 4.1 and Figure 4.11 present the dependencies identified in Figures 4.5 to 4.10.

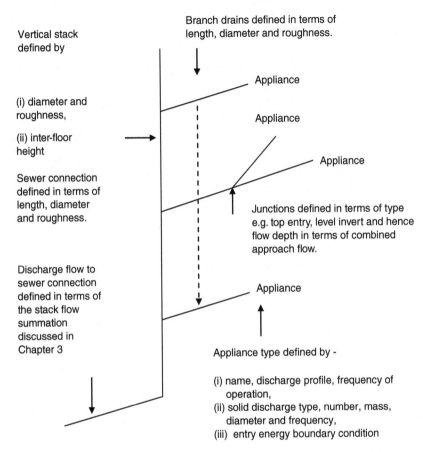

Figure 4.3 Schematic of DRAINET graphical user interface that allows multi-storey systems to be modelled.

The roughness parameter was investigated by reference to the simulation of flow in four particular drain pipe materials, uncoated cast iron, coated cast iron, UPVC and glass; the roughness values are presented in Figure 4.2.

Figures 4.5 to 4.10 illustrate clearly the mechanism of wave attenuation driven by the variation of wave speed along the wave profile. The steepening of the leading edge is demonstrated; however the action of frictional forces prevents all the surges from becoming steep fronted waves that 'break' at some point along the channel. In addition to decreases in peak depth, Figures 4.5 to 4.10 also illustrate the effect of channel slope and friction on the propagation speed of the wave as for a steeper slope and a smoother surface the leading edges of the waves in the steeper and smoother channels outpace those in milder slope or rougher channels – this is particularly clear in Figure 4.7 where the effect of pipe slope is simulated.

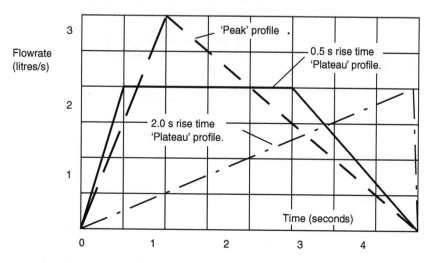

Figure 4.4 Illustrative discharge profiles to demonstrate dependence of wave attenuation on channel and appliance discharge parameters.

Table 4.1 Summary of the dependence of wave attenuation on drain and flow parameters

Parameter	Figure	Attenuation outcome
Drain diameter	4.5	Increases as diameter is reduced
Drain roughness	4.6	Increases with wall roughness
Drain slope	4.7	Increases as drain slope reduced
Discharge profile	4.8	Peak profiles display greater attenuation.
Discharge profile rise time	4.9	Increases with steepening of discharge leading edge
Baseflow	4.10	Decreases as baseflow increased

Table 4.2 Wall roughness values used in the determination of an appropriate friction factor within DRAINET

Drain pipe material	Roughness, mm, used in the Colebrook-White friction factor determination
Uncoated cast iron	0.30 mm
Coated cast iron	0.15 mm
UPVC	0.06 mm
Glass	0.03 mm

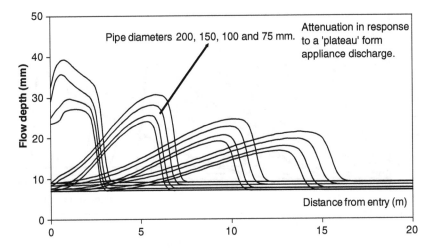

Figure 4.5 Influence of pipe diameter on the attenuation in a 20 m, 100 mm
diameter glass drain at a slope of 1/100 in response to a 'Plateau' format
appliance discharge (Figure 4.4).

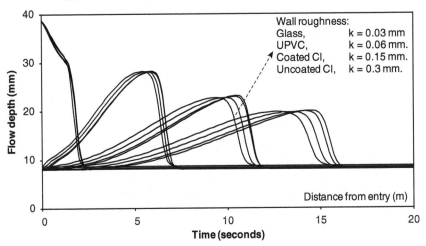

Figure 4.6 Influence of wall roughness on the attenuation in a 20 m, 100 mm
diameter drain at a slope of 1/100 in response subject to a 'Plateau'
format appliance discharge (Figure 4.4). Pipe materials include coated
cast iron, uncoated cast iron, UPVC and glass.

While Figures 4.5 to 4.10 have concentrated on flow depth attenuation,
the DRAINET simulation also allows the prediction of other relevant flow
parameters, including flowrate, flow velocity, wave speed, specific energy, a
combination $h+0.5\rho u^2$, and Froude Number, the ratio u/c. Figures 4.12, 4.13
and 4.14 illustrate the wave attenuation over a 20 m long pipe length via the
peak profile for each of these parameters.

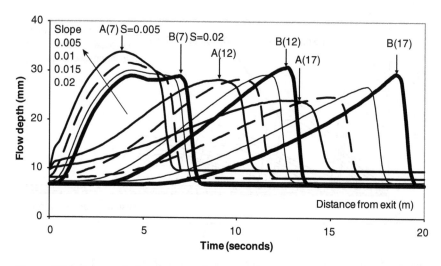

Figure 4.7 Influence of pipe slope on the attenuation in a 20 m, 100 mm diameter glass drain at slopes of 0.02, 0.015, 0.01 and 0.005 in response subject to a 'Plateau' format appliance discharge (Figure 4.4).

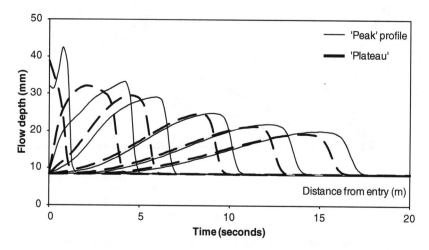

Figure 4.8 Influence of appliance discharge profile on the attenuation in a 20 m, 100 mm diameter coated cast iron drain at a slope of 1/100 in response subject to either a 'Plateau' or 'Peak' format appliance discharge (Figure 4.4).

4.2 Observation of wave attenuation and its impact on drain network capacity

The simulations undertaken illustrate clearly that wave attenuation implies a higher drain capacity for a given discharge profile than that recommended

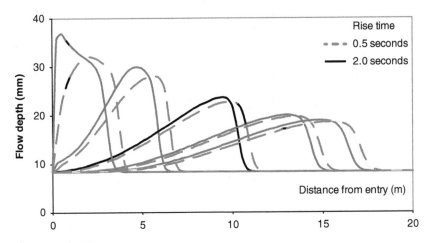

Figure 4.9 Influence of leading edge rise time on the attenuation in a 20 m, 100 mm diameter coated cast iron drain at a slope of 0.01 in response subject to a 'Plateau' format appliance discharge (Figure 4.4) with rise times of 0.5 and 2.0 seconds.

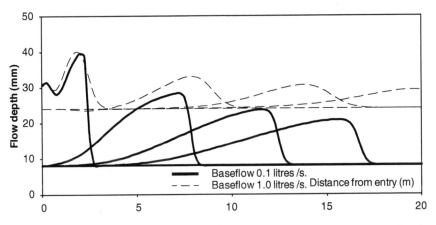

Figure 4.10 Influence of baseflow on the attenuation of a 'Peak' profile format discharge in a 20 m, 100 mm diameter glass drain at a slope of 0.01 in response subject to a 'Plateau' format appliance discharge (Figure 4.4).

by current codes based on steady flow theory, as illustrated in Chapter 2. Burberry (1978) indicated that for drains in excess of 10 m attenuation reduced the discharge peak to below half its original level. This observation of course only applies to appliance discharges with relatively short durations and displaying a peaked profile – i.e. it applies to w.c. discharges but not to

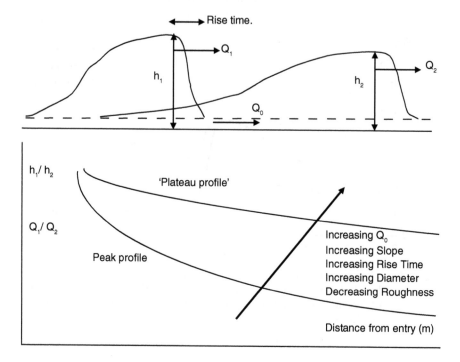

Figure 4.11 Summary of attenuation dependencies.

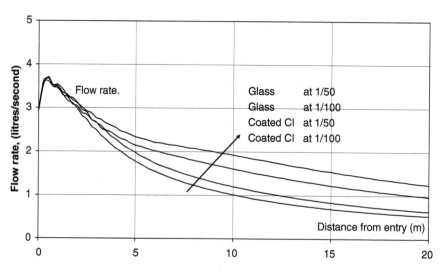

Figure 4.12 Peak flowrate profiles along a 20 m long, 100 mm diameter drain in response to a 'Peak' profile discharge. The illustrated cases represent coated cast iron and glass branch drains set at slopes of 1/100 and 1/50.

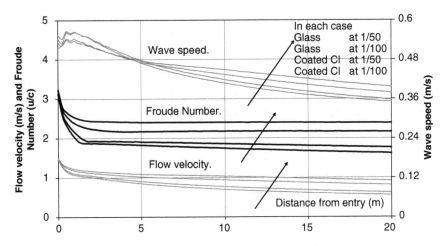

Figure 4.13 Peak flow velocity, wave speed and Froude Number profiles along a 20 m long, 100 mm diameter drain in response to a 'Peak' profile discharge. The illustrated cases represent coated cast iron and glass branch drains set at slopes of 1/100 and 1/50.

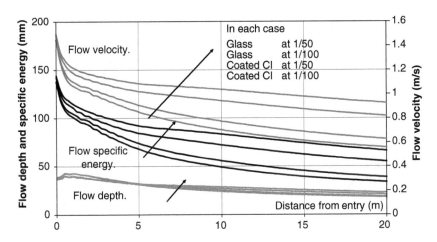

Figure 4.14 Peak flow velocity, depth and specific energy profiles along a 20 m long, 100 mm diameter drain in response to a 'Peak' profile discharge. The illustrated cases represent coated cast iron and glass branch drains set at slopes of 1/100 and 1/50.

baths, showers etc. However the peak discharge flow from w.c.'s is much greater than the longer time base discharges from other appliances so the point is wholly relevant.

Burberry (1978) presents data for the observed time taken for a w.c. discharge wave to pass a series of monitoring points along a drain set at a range of slopes from 1/40 to 1/120. Unfortunately the paper is not explicit as to the instrumentation used, and it is assumed that the time duration of the flow was simply recorded with a stopwatch. Experience suggests that this is a relatively crude approach and that the time duration figures may be less than accurate as the trailing edge of a wave is difficult to identify, and hence the times may be approximate. Similarly it is unlikely that the duration of passage of the wave would be a linear function of distance from entry. However the results are important as they represent one of the earliest attempts to quantify the effect of wave attenuation. Figure 4.15 presents comparisons to the DRAINET simulation for the 1/40 and 1/80 cases, while Figure 4.16 illustrates the increasing predicted wavelength for the appliance discharge propagating along the drain set at a slope of 1/140.

In addition Burberry (1978) presents a curve of likely peak flowrate for a w.c. discharge, and the resulting values of peak flow at a series of locations along the drain are again compared to the DRAINET predictions (Figure 4.15). Note that the flowrate data presented is the ratio of flowrate to peak flowrate (DRAINET) and to the peak w.c. flow allocated by the drainage codes appropriate in 1978 (Burberry).

The importance of wave attenuation in determining the capacity of horizontal drains was also appreciated by Wyly (1964), who presented a series of capacity curves based on experimental work at NBS Washington, DC. Figure 4.17 illustrates Wyly's observations compared to a DRAINET simulation (Swaffield and Galowin 1992). The characteristic reduction in peak flowrate is again clearly demonstrated and introduces the concept that drain capacity is a function of the applied discharge profile, the distance the wave has travelled and the channel parameters of diameter, roughness and slope.

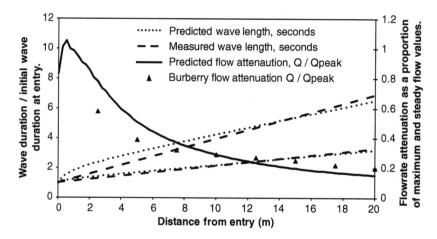

Figure 4.15 Comparison of DRAINET predictions of wave attenuation, in terms of flowrate and wavelength, to the measurements presented by Burberry (1974).

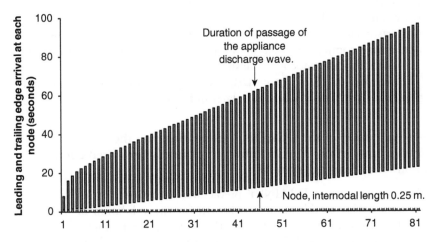

Figure 4.16 Graphical representation of the predicted increasing length of the appliance discharge as it progresses along a 20 m, 100 mm diameter drain set at a slope of 1/140. The time taken to pass each node represents the increasing physical length of the wave.

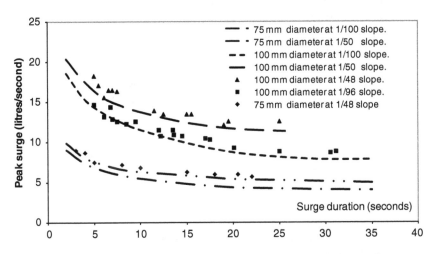

Figure 4.17 Comparison of Wyly (1964) observation of horizontal drain flow surge capacity with DRAINET simulations (Swaffield and Galowin 1992).

The experimental work reported by both Wyly and later Burberry confirm that wave attenuation offers an opportunity to reconsider steady-state drainage load calculations. The reliance on the Hunter (1940) model of flow assessment is therefore challenged; however there are clear indications in Hunter's seminal work that he recognised the limitations of his approach – limits imposed by a lack of analysis capability – a capability now readily available in the form of accessible fast computing support. The comparisons presented also provide a degree of validation for the DRAINET simulation – comparisons that are particularly valuable as the observational data is wholly independent of the simulation development process.

During the development of DRAINET specific validations were undertaken and these will be referred to later in this chapter.

4.3 Validation of the DRAINET simulation and the applicability of the MoC solution to unsteady flow prediction in building drainage systems

The comparisons of DRAINET predictions to historic data (Wyly 1964; Burberry 1978) establishes a benchmark for the validation of the simulation based on the MoC solution of the St Venant equations of continuity and momentum. In addition direct validation was undertaken for the entry boundary conditions already discussed in Chapter 3, namely the energy-related boundary to represent high-velocity discharge from a w.c. appliance and the more tranquil Normal or Critical depth entry conditions associated with baths, showers, sinks etc. as well as the Critical depth boundary immediately downstream of a junction or flow obstruction.

Figure 4.18 illustrates an experimental arrangement at Brunel University (Swaffield and Galowin 1992) where the w.c. discharge is replicated by full bore flow to the head of the drain via a small-diameter vertical discharge pipe, the boundary condition already introduced in Chapter 3. The agreement between the DRAINET predictions and the measured flow depths is satisfactory; the attenuation of the wave is also accurately predicted.

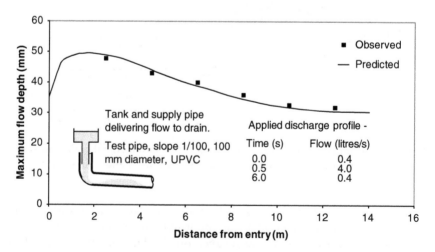

Figure 4.18 Validation of the energy entry condition representing a w.c. discharge to a horizontal branch drain.

Figure 4.19 presents a similar validation to ensure that the tranquil flow entry to a horizontal branch from such appliances as baths, showers and sinks is also adequately modelled. The comparisons presented indicate that again both the flow depths and wave attenuation are adequately modelled.

In order to simulate the operation of a vertical stack in delivering upper floor discharges to a sewer connection it is necessary to model the summation of individual floor discharges to the stack in order to drive the flow conditions and wave attenuation in the lowest level sewer connection, as shown in Figure 4.20, illustrating a plumbing tower installation at NBS Washington, DC. Figure 4.21 demonstrates the agreement between observed and predicted flow depth 6.5 m along the 75 mm diameter UPVC drain, set at a slope of 0.06. As with the earlier validation the results indicate a satisfactory application of the MoC solution, as well as providing confidence in the numerical stack flow summation presented in Chapter 3.

4.4 Wave attenuation as a result of stack flow to a horizontal sewer connection or multi-stack building drainage network collection network

The application of simulations, such as DRAINET, allows the investigation of phenomena not readily addressed in the laboratory due to the range of parameters affecting the overall outcome. The transition from stack annular downflow to horizontal branch or sewer connection free surface flow is one such application that has implications for the design of drainage networks.

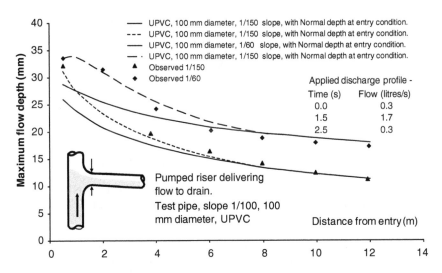

Figure 4.19 Validation of the Normal and Critical depth boundary conditions representing the more tranquil discharge to a horizontal branch drain from appliances such as baths, showers or downstream of a junction or defect.

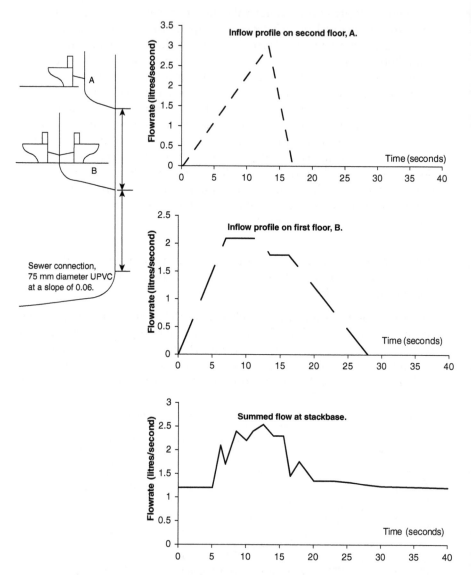

Figure 4.20 Summed flow at the base of a two-storey vertical stack at NBS Washington, DC (Source: Swaffield and Galowin 1992).

Historically the transition from stack to horizontal branch has not been well understood and has in a way encapsulated the difficulty that drainage flows have in some way been seen as 'different', rather than being just a particular example of steady and unsteady fluid mechanics.

In addition the simulation of the stackbase transition will draw together several issues already addressed, for example the influence of stack diameter

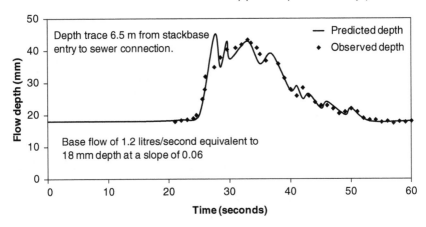

Figure 4.21 Comparison of flow depth – the observed and predicted flow depths downstream of the vertical stack to sewer connection, NBS test installation (Figure 4.20).

and roughness on the annular flow and the flow terminal velocity; the importance of the assumptions made as to the energy transfer at the stackbase and hence the initial entry energy possessed by the free surface flow; the importance of the downstream horizontal branch diameter, roughness and slope on the subsequent wave attenuation and depth profiles downstream and the influence of the upper floor appliance operation sequence and loading on the stack.

Figure 4.22 illustrates the multi-storey network to be used in the simulations presented in this section. The system consists of four upper floors and a collection horizontal drain at the lowest level representing a sewer connection or a horizontal collection drain in a multi-stack network. In the latter case the inflow profile to the horizontal branch from the stack illustrated can be treated as an equivalent appliance discharge profile and the collection network at whatever level it exists within the building treated as a separate network, illustrating the capability of the DRAINET simulation to model complex buildings by building up stored profile data representing any stack vertical configuration.

The network on each upper floor is represented by a single w.c. appliance located 2 m from the stack so that attenuation in the upper floor branches is effectively eliminated from the simulation. This is acceptable as any attenuation would reduce the peak loading on the stack and hence generate lower level results less challenging than those to be demonstrated here.

The stack, horizontal branch and collection drain will initially be of 100 mm diameter, with the horizontal drains set at a slope of 1/100. The drain and stack are initially glass or uncoated cast iron – representing the opposite ends of the roughness spectrum, $k = 0.03$ and 0.3 mm respectively in the Colebrook-White friction factor expression.

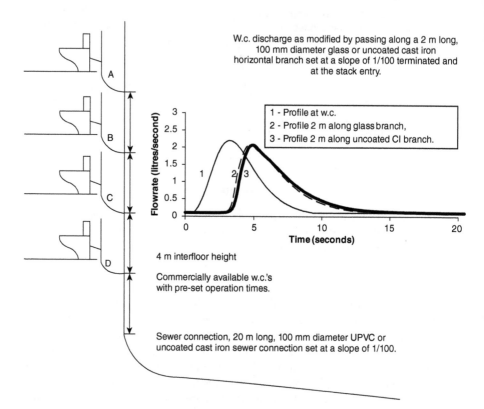

Figure 4.22 Schematic of the vertical stack and sewer connection network used to demonstrate the DRAINET simulation of vertical stack entry flow to a horizontal sewer connection or multi-stack collection network.

The sequence of upper floor appliance discharge times will also be a parameter, along with the time taken for the discharge to reach the stack and then fall vertically to the stackbase – interacting with any lower level inflows en route. It will be demonstrated that simultaneous discharge is not the worst case. The worst case will be network specific, dependent upon arrival time at the stack and travel time down the stack, itself dependent upon stack terminal velocity based on stack diameter and roughness.

The following simulations will therefore demonstrate the influence of these parameters for the specific network illustrated in Figure 4.22.

Figure 4.23 illustrates the accumulated discharges from the upper floors at the stackbase to horizontal drain interface. Two cases are presented for the glass network, namely simultaneous discharge on upper floors and a 1 second delay between floors, i.e. the uppermost floor w.c. discharges 3 seconds ahead of the lowest level w.c. This timing offset has the effect of enhancing the inflow profile at the stackbase as it allows the upper floors to arrive at roughly the same time as the lower floor discharges. It is of course evident that this is wholly network

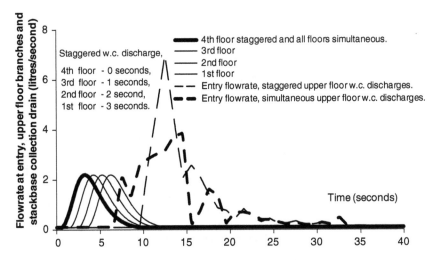

Figure 4.23 Comparison of the horizontal sewer connection drain entry flow profile following simultaneous and staggered w.c. discharges on the four upper floors. All branches and vertical stack 100 mm diameter glass, branches set at 0.01 slope.

specific, and a reference to probability theory will demonstrate that such congruity will be unusual, dependent upon the number of upper floors, the number of appliances on each floor, the duration of the discharge at each appliance and the frequency of use of each. However the cases presented do indicate a higher peak inflow and a more unified profile compared to the multi-peak profile due to the simultaneous discharge of the upper floor appliances.

Figure 4.24 illustrates the maximum depth and flowrate profiles along the horizontal drain as a result of the simultaneous and staggered inflow profiles presented in Figure 4.23. The influence of the stackbase entry boundary condition is immediately obvious. In Chapter 3 the stackbase energy boundary condition was established by reference to the kinetic energy of the annular flow transiting into free surface horizontal branch flow. A simple assessment of likely 'energy loss' at the stackbase – actually energy transfer due to the flow disruption at the stackbase transition – identified a boundary condition where 50% of the annular flow kinetic energy was available to be transferred into horizontal branch entry specific energy – based on a combination of horizontal branch flow depth and kinetic energy. This results in the branch entry condition being characterised as shallow fast flow, with values below either the Normal or Critical depths for the applied flow and drain parameters. Therefore the free surface flow initially increases rapidly in depth and reduces in velocity before wave attenuation becomes the defining mechanism. This results in the peak profiles presented in Figure 4.23.

Figure 4.25 demonstrates the changing flow profile along the length of the drain at a series of defined times post stackbase inflow increase for the

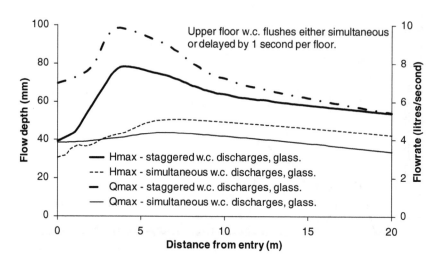

Figure 4.24 Comparison of the maximum flow depth and flowrate profiles along the 20 m, 100 mm diameter glass sewer connection branch set at 0.01 slope.

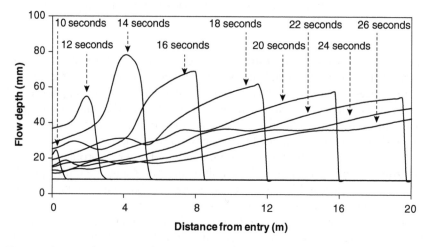

Figure 4.25 Development and attenuation of the discharge wave along the 20 m, 100 mm diameter glass sewer connection at a 0.01 slope.

staggered appliance discharge case – chosen as the inflow rate and subsequent flow depths were greater. These 'freeze frame' views of the wave progression along the drain are taken from the data utilised to generate the REALTIME capability of the DRAINET simulation to present wave motion and attenuation throughout a given network, a valuable initial view of whether a network has design problems in terms of flow capacity.

Figure 4.26 presents complementary data for an uncoated cast iron network. It will be seen that the appliance discharges are in this example set at a delay sequence of 1 second per floor and 1.45 seconds per floor. The

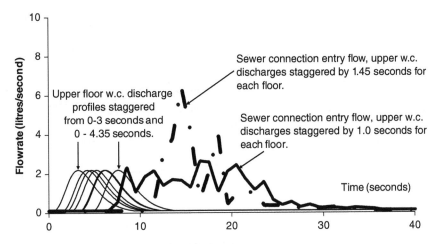

Figure 4.26 Comparison of the horizontal sewer connection drain entry flow profile following 1.0 and 1.45 second increment staggered w.c. discharges on the four upper floors. All branches and vertical stack 100 mm diameter uncoated cast iron, branches set at 0.01 slope.

difference between the glass and uncoated cast iron 1 second delay case – namely a much reduced peak flow profile for the uncoated cast iron network – is explained in terms of the lower stack annular flow terminal velocity that has a key role in developing the combined flow profile at the base of the stack, as discussed in the development of the stack flow summation mechanism in Chapter 3. In order to generate a more unified stackbase profile an increased delay is used as shown. A further increase in delay time to 2 seconds per floor will be discussed later with reference to possible full bore flow onset in the horizontal drain.

Figure 4.27 presents a comparison of the 1 second delay appliance discharge case for both the glass and uncoated cast iron applications of DRAINET. As expected from a review of the inflow profiles the cast iron case displays much reduced peak flowrates and depths, although the form of the relationships are identical for both cases.

Figure 4.26 illustrated a twin peak inflow profile for the 1.45 second delay case and this has an impact upon the form of the downstream flow depth profiles along the horizontal drain, as illustrated in Figure 4.28. The second inflow peak generates a second wave that 'runs' along the preceding flow depth profile. As the flow depth is greater it will be seen from Figure 4.28 that this second wave 'catches up' the leading edge of the flow profile, resulting eventually in a steep-fronted wave with the capability to travel considerable distances along the drain as a wave front. This observation will be referred to later in a discussion of tipping tank and siphon tank operation as flow increasers to offset perceived solid deposition problems as a result of reduced water usage.

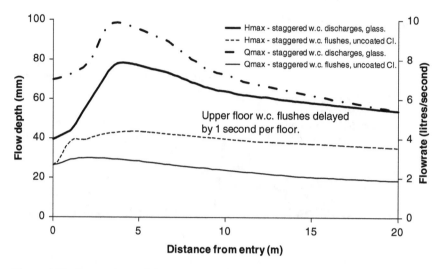

Figure 4.27 Comparison of the maximum flow depth and flowrate profiles along the 20 m, 100 mm diameter glass and uncoated cast iron sewer connection branches set at 0.01 slope.

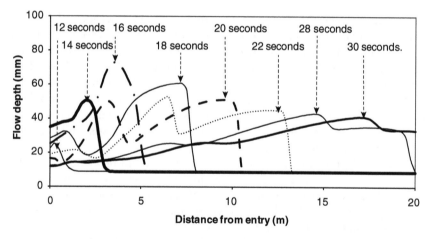

Figure 4.28 Development and attenuation of the discharge wave along the 20 m, 100 mm diameter uncoated cast iron sewer connection at a 0.01 slope, following the 1.45 second staggered upper floor w.c. discharges. Note the second peak wave generated by the double peak inflow profile (Figure 4.26) and the relative propagation velocities of the first and second waves.

As expected from these observations, Figure 4.29 indicates that the 1.45 second delay applied to upper floor appliance discharges generates a greater maximum flow depth and flowrate profile for the uncoated cast iron network. It is also noticeable that there is a greater attenuation in the peak

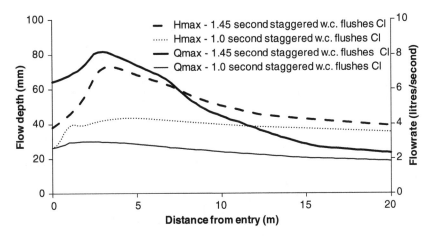

Figure 4.29 Comparison of the maximum flow depth and flowrate profiles along the 20 m, 100 mm diameter uncoated cast iron sewer connection branch set at 0.01 slope.

flowrate generated by the 1.45 second delay simulation, explained by the influence of inflow profile on attenuation already discussed in this chapter; the 1.45 second delay inflow has a higher peak compared to the 1.0 second delay profile and is steeper, influences already identified in terms of the 'peak and plateau' inflow profile descriptors used previously.

Figure 4.30 compares the stackbase flow profile summations already discussed and includes a further cast iron network case where the upper floor w.c. delay interval has been increased to 2 seconds. The data presented in Figure 4.30 has been modified by adding a 20 second start time increment to all the cast iron results to differentiate more clearly the various curves. It will be seen that increasing further the cast iron delay has emphasised the single peak format for the summed stackbase to horizontal drain inflow profile, while retaining the overall shape of the profile.

Figure 4.31 reproduces the subsequent maximum flow depth and flowrate profiles predicted for the 100 mm and 150 mm diameter uncoated cast iron horizontal drains. It will become immediately apparent that the downstream drain goes full bore within 10 seconds of the arrival of the surge profile from the upper floors. Clearly this is a design failure; one of the objectives of the DRAINET simulation was to identify such failures and to recommend design modifications. In practice using DRAINET in this way and then accessing the automatic REALTIME function to display wave propagation along the horizontal network would immediately identify the problem and suggest an immediate solution – namely an increase in horizontal drain diameter. In theory increasing the drain diameter immediately downstream of the stack

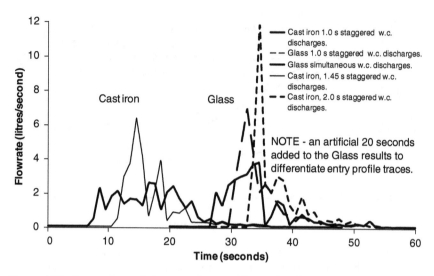

Figure 4.30 Comparison of the entry flow profiles to the glass and uncoated cast iron sewer connections. Note profiles artificially separated for clarity.

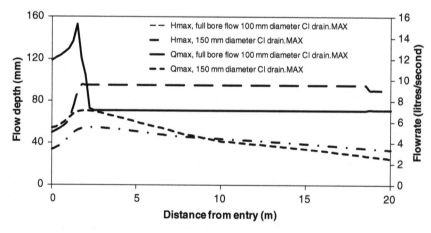

Figure 4.31 Comparison of the maximum flow depth and flowrate along a 20 m sewer connection of 100 mm and 150 mm diameter at a slope of 0.01.

base to horizontal drain interface would be sufficient as the subsequent wave attenuation would allow a smaller diameter downstream section to cope with the surge, but reductions in drain diameter downstream is specifically disallowed in all national codes; however careful design of the stackbase to horizontal drain interface would assist as sweeping the entry bend could reduce the energy 'loss' thereby extending the development length and allowing more scope for the natural wave attenuation to reduce flow depths.

Figure 4.31 illustrates the effect of increasing the cast iron network diameters to 150 mm from the initial 100 mm. Note this increase is applied to the stack as well as the horizontal collection drain but not to the upper floor branches delivering the w.c. discharges to the stack so that changes in annular terminal velocity also affect the outcome. Figure 4.32 illustrates the subsequent decrease in inflow peak loading on the horizontal sewer connection, in this specific case from 12 litres/second to less than 6 litres/second. This has the desired effect on downstream conditions, as illustrated by Figure 4.33 where the maximum flow depth is well below 50% of the

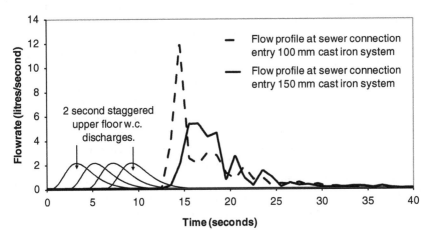

Figure 4.32 Comparison of the entry flow profile generated by upper floor w.c. discharges staggered by 2 seconds in a 100 mm and a 150 mm diameter sewer connection.

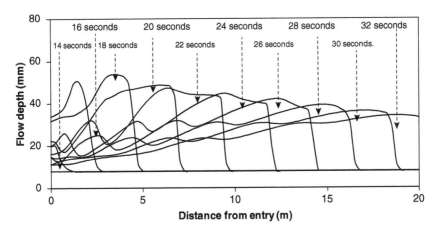

Figure 4.33 Wave attenuation along the 20 m, 150 mm, diameter uncoated cast iron sewer connection following the 2.0 second staggered set of upper floor w.c. discharges.

drain diameter and therefore acceptable within most national codes. The effect of drain diameter increase is also demonstrated clearly by Figure 4.34 where the flow depth profiles along the drain are compared for the two diameter cases.

The onset of full bore flow also raises an observational validation for the MoC simulation. When full bore flow is initiated it is seen to 'flash' along the drain from inception at the first location to experience full bore conditions as the flowrate rises. The explanation for this is given by the wave speed expression $c = \sqrt{gA/T}$. When the flow depth → full bore depth D the value of c tends to rise sharply and hence the wave accelerates along the drain, extending the full bore flow rapidly, as illustrated in Figure 2.3. Figure 4.35 illustrates this effect for the 100 mm diameter uncoated cast iron case with a 2 second appliance discharge delay applied to the upper floors.

Reference to Figure 4.35, which illustrates the establishment of full bore flow, indicates that the condition is reached at 2 m into the drain at 15 seconds into the flow event. The full bore flow condition is then propagated along the drain, reaching the 18 m point by 30 seconds into the event. The average propagation speed of the wave is thus around 1.6 m/s. Note that in order to allow the simulation to continue to demonstrate this effect the onset of full bore flow is defined as a flow depth in excess of 95% of drain diameter. The full bore flow depth is then held at this value. DRAINET will also simply identify the full bore flow condition and cease the modelling at that time. Absolute accuracy post full bore flow establishment is not required as it is sufficient to identify the location of the system failure. Figure 4.36

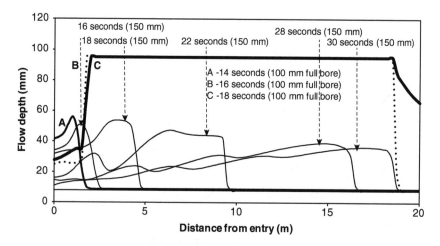

Figure 4.34 Demonstration of the effect of increasing the horizontal sewer connection diameter from 100 mm to 150 mm on the full bore flow established following a series of 2 second delay upper floor w.c. discharges. Note the maximum depth is less than 50% of drain diameter.

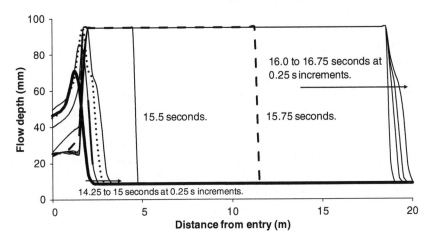

Figure 4.35 Propagation of the full bore flow condition along the 100 mm uncoated cast iron drain in response to the inflow profile generated by upper floor w.c. discharges at a 2 second stagger. Note full bore flow first reached at 2 m from entry at 15 seconds into the event.

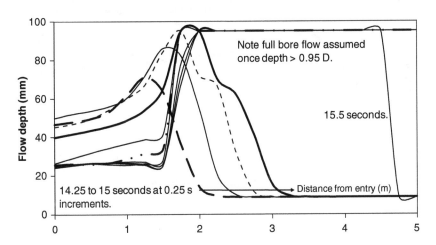

Figure 4.36 Initial propagation of the full bore flow condition along the 100 mm uncoated cast iron drain. As the inflow wave loses momentum its depth increases until a full bore condition is reached 2 m from entry at 15 seconds into the event.

presents more detailed depth development traces close to the drain entry and reinforces the mechanism of full bore flow establishment and the rapid increase in wave propagation velocity once the flow depth exceeds 0.95D, as illustrated by the separation of the 15 and 15.5 second traces.

The DRAINET simulations of sewer connection or collection drain operation also clarify a long-standing erroneous definition of the flow conditions downstream of a stack to horizontal drain. It was common to find the flow condition downstream of the stackbase interface described by the expectation of a 'hydraulic jump' formed in the horizontal drain leading to a potential full bore flow situation, for example Orloski and Wyly (1978). Figure 4.37 illustrates the definition presented.

It has already been shown in Chapter 2 that a hydraulic jump may only occur if there is a rapidly varying flow transition from supercritical to subcritical flow conditions, as is the case upstream of a junction or an obstruction in a normally supercritical flow – a case that encompasses the vast majority of building drainage system installations as opposed to sewers. Chapter 2 identified that downstream of a stack entry a gradually varied flow profile would be established under steady flow conditions that transits the flow from its entry depth to the normal depth appropriate to the drain parameters of slope, diameter and roughness.

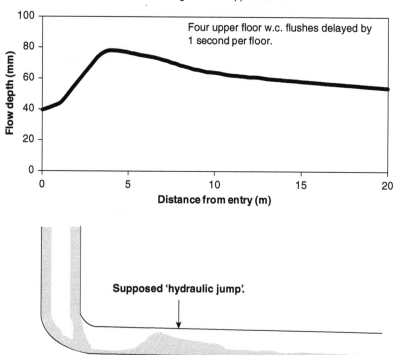

Figure 4.37 Historic definition of flow conditions downstream of a stack to collection drain interface (Orloski and Wyly 1978) compared to the flow profile predicted by DRAINET for the 100 mm diameter uncoated cast iron drain (Figure 4.27).

Figure 4.37 illustrates the maximum depth profile along the collection drain at a slope of 1/100 in response to the upper floor w.c. discharge sequence. It will be seen that a maximum depth is predicted close to the drain entry. The explanation is that the shallow fast entry flow decelerates rapidly in the first few metres of the drain and therefore the flow depth increases rapidly. However at the same time the wave attenuation mechanisms act to reduce the flow depth downstream and extend the wave length as fully discussed in this chapter. The resulting flow profile, the profile that would be observed in the drain, would have led to the erroneous definition of the flow condition. The ability of the MoC solution to the St Venant equations of continuity and momentum, coupled to the development of a rational boundary condition to describe the stack to drain interface, has allowed this clarification.

The MoC-based simulation of free surface transient flows in building drainage systems has been particularly successful in predicting the flow conditions downstream of a stack to collection drain interface. As such it meets the requirement in Orloski and Wyly (1978), a requirement published some 3 years prior to Swaffield and Galowin's introduction of the technique, (Swaffield 1981b):

> development of insight into and understanding of this phenomenon (the hydraulic jump mechanism) in plumbing systems is important to the evaluation of innovative systems . . . A more general definition of the conditions determining the drain capacity is required.
>
> Orloski and Wyly (1978) pp. 12–13

4.5 Simulation of junction effects within horizontal branch free surface flows

In order to model a complex building drainage network and its response to random appliance discharges on several floors it is necessary to include in the simulation representation of the confluence of flows at horizontal system junctions and the flow summation that takes place between floors in the system's vertical stacks.

The simulation of the vertical stack flow summation and the response of the downstream collection drain have been demonstrated and have confirmed the application of the MoC methodology to determining the flow capacity of such collection drains under unsteady flow conditions.

It is also necessary to detail the response of the flow conditions around a horizontal system junction in order to simulate the response of the network to multiple appliance discharges and, in particular, to demonstrate the mechanisms of backflow at such junctions. Chapter 3 detailed the boundary conditions required at such junctions, and these will be deployed in the following simulations to demonstrate the mechanisms of junction–flow interaction.

Bridge (1984) presented the initial MoC simulation conditions based on a survey of relevant junction analysis. The fundamental characteristic of junction operation is that the flow depth at the junction is common for all approach branches. In addition the discharge depth to the downstream branch may be characterised by the flow passing through a Critical depth location immediately downstream of the junction. The definition of a common junction depth allows the initial steady flow conditions in all the joining branches to be defined and, in the normal drainage case that all the approach branches will normally be carrying supercritical flow, allows the initial position of the upstream hydraulic jumps to be determined. The location of these jumps relative to the junction will, as demonstrated in Chapter 2, be governed by the branch diameter, roughness, slope and the initially steady baseflow assumed to be applied to the network.

The arrival of the appliance discharges along one or more of the approach branches will result in the jump locations responding to the changed flow conditions. The jumps will move upstream in response to a rising hydrograph in its branch or to a rising hydrograph in any one of the other branches joining at the junction. This latter effect is due to the common junction depth increasing and hence generating a reverse flow condition in all the other joining branches. As the approach flow in any branch rises further the jump continues to move upstream; however on cessation or reduction in approach flow it will move back towards the junction, reverting to its start position if the flow conditions revert to the initial steady state baseflow condition.

In an analysis of acceptable iteration techniques, Vardy (1990) stresses the importance of the simulation conditions being able to return automatically to the start condition at the culmination of the unsteady flow event – the

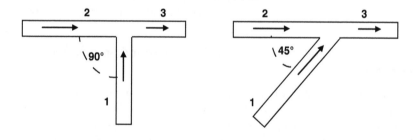

Level invert junctions viewed in the horizontal plane.

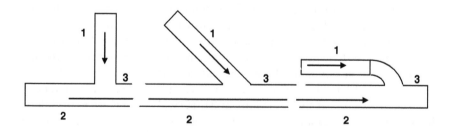

Top entry junctions viewed in the vertical plane.

Figure 4.38 Level invert and top entry junction geometry utilised in the demonstrations of MoC simulations of junction–flow interaction.

demonstrations already presented all indication that this is achievable and this will also become apparent in all the junction–flow interactions that follow.

Two categories of junction will be discussed, namely level invert junctions, with either 90° or 135° geometry, and top entry junctions where the upper supply branch is either vertical, inclined at 45° to the horizontal, or is defined as being 'in-line' with the horizontal branch, entry to the lower branch being accomplished via a 90° bend in the vertical plane. Figure 4.38 illustrates these schematic options.

With level invert junctions, the discharges from the joined pipes impinge upon each other within the junctions, causing the flow depth in each pipe to increase, and the hydraulic jump and backwater profiles exist in each pipe if the flow in each pipe leading to the junction is supercritical. The effect of the junction is to impede the flow and cause a subcritical backwater profile to develop. This phenomenon is illustrated in Figure 4.39. The graph shows the flow along pipes 2 and 3, a total of 12 metres. The junction is located at 10 metres from entry, and the characteristics of combining flows from pipes 1 and 2 influence the hydraulic jump location and magnitude. It can be seen in this instantaneous snapshot that the water depth rises at 8.75 m from entry and that the effect of this rapidly varying flow is to cause a sudden depression in velocity. The graph shows minimum and maximum values for water depth and velocity to illustrate the ability of DRAINET to predict the changing backwater profile and position of the hydraulic jump when the system is subjected to an unsteady discharge.

While Figure 4.39 shows location, depth and velocity information along the whole length of the pipe, from inlet to outlet, the interaction of flows from pipe 1 and its influence on the flow in pipe 3 is shown in Figure 4.40.

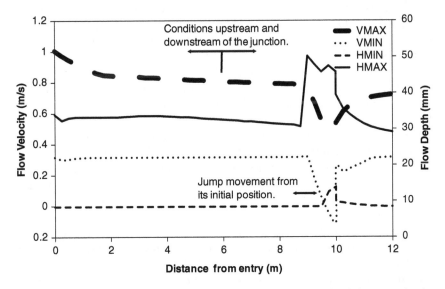

Figure 4.39 Maximum and minimum flow velocity and water depth for a 90° level invert branch junction.

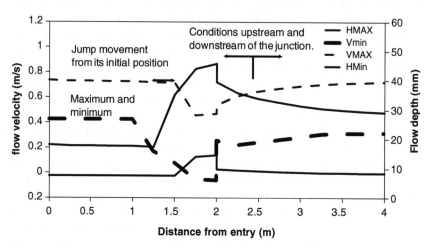

Figure 4.40 Interaction of flows from a 90° branch junction.

This interaction is characterised by a slightly negative flow velocity due to the impediment caused by the confluence of flows and the moving subcritical backwater profile in the hydraulic jump. The ability of DRAINET to predict such complex interactions at junctions allows a qualitative approach to be taken in this area of system operation where much attention will inevitably focus in a reduced water usage context, since historically, junctions are more prone to blockages than any other part of a horizontal drainage system.

These conditions change when the location of the inlet of the junction changes from 90° level invert to 45° level invert. This change is illustrated in Figures 4.41 and 4.42 for the same simulation scenario, subjected to the same unsteady flows.

There are considerable similarities between the prediction of these two level invert options. It will be shown later that both of these junction types present challenges for clearing solids travelling in the pipe, being conveyed by the water, and that risk of blockages is considered increased, particularly under water conservation and reduced flow criteria. The interaction of flows for the 45° level invert junction is shown in Figure 4.42 also.

While level invert junctions are the most common fittings in a near horizontal network, top entry junctions are also an option. With top entry junctions, the branch flow enters the main pipe from above, and so only the action of the branch-pipe discharge presents an obstruction to the main pipe flow of the branch which is considered as a free discharge, causing a reduction in velocity and an increase in flow depth upstream of the point of confluence. There is no effect on the branch flow from the main pipe flow, in a partially filled pipe, free surface wave context.

Figure 4.43 illustrates how the hydraulic jump develops with time at the junction between the branch and the main collection drain line. It can be seen

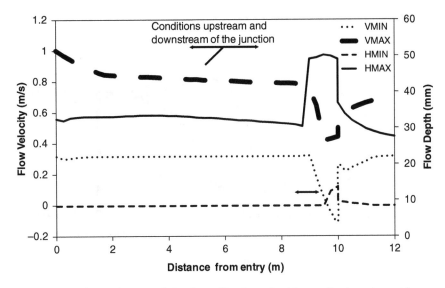

Figure 4.41 Flow velocity and depth profile along the 12 m collection pipe with a 45° level invert junction at 10 m.

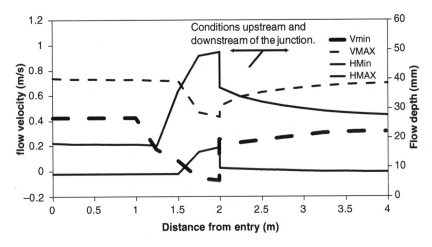

Figure 4.42 Interaction of flows at a 45° level invert junction.

how the relatively shallow depth of 20 mm appears at the junction located at 10 m from pipe entry at 1 second into the simulation. It can also be seen that the water depth rises and moves backwards over time until the main peak depth of 44 mm occurs at 8.5 m from entry, 5 seconds into the simulation. The jump 'holds' position and depth until it begins to collapse and move back towards the junction at 10 m at 9 seconds into the simulation. This ability to track the temporal and spatial movement of water depth and velocity provides both

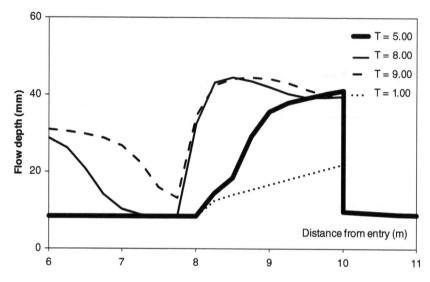

Figure 4.43 Flow depths along the main collection drain (pipes 2 and 3)
illustrating development of the jump position at different times for a
top entry junction.

qualitative and quantitative data about the operation of this junction. While the
hydraulic jump in this case is significant, at almost half the diameter of the pipe,
there is no effect on the adjoining branch, and the probability of blockage due
to leakage of material into the branch is greatly reduced.

The depth and velocity minimum and maximum values at the interacting
junction are illustrated in Figure 4.44.

The situation for a 45° top entry junction flows a similar pattern; however
the inlet flow is smoother in the direction of flow in the main drain line. The
velocity and flow depth profiles are shown in Figure 4.45 and Figure 4.46.

Note the much less pronounced flow depth profile in Figure 4.45, indicating
that there is much less interaction between the flows, and that the discharging
branch is less disruptive than the 90° inlet or the level invert junction already discussed. Note also the rapid reduction in velocity in Figure 4.46 and the upstream
movement of the backwater profile of the hydraulic jump as a result of the rapidly
varying flow initiated by the unsteady flow discharge from the top entry branch.

The options shown for branch inlet give varying degrees of protection against
the risk of blockages occurring in the main drainline or in the branch itself. The
modelling of these effects and other flow phenomena described in this chapter
offer a powerful tool with which to evaluate changes in usage patterns and
technical approaches to design and compliance with standards. Concerns raised
over the likely failure of some systems under low water usage and water conservation strategies can easily be tested without having to build elaborate test rigs.

This concern over reduced water usage, a direct response to calls for
water conservation and saving precious resources, has led some researchers

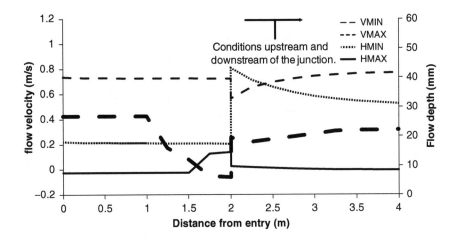

Figure 4.44 Interaction of flows for top entry junction.

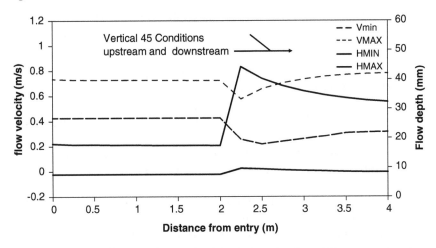

Figure 4.45 Flow velocity and water depth at the junction for a 45° top entry junction.

to focus on the effects on the safe operation of the drainage system under extreme usage conditions. An ability to model junction interactions and the effects of different branch junction types is crucial in understanding how possible blockages can occur, and more importantly, how to design them out of systems as much as possible. In addition to this there is a need to be able to model solid transport and the expected loading of obstructions. This will be dealt with in depth in Chapter 5; however mention is made here of this aspect of system performance to illustrate its importance.

The graphs in Figures 4.39 to 4.46 show interactions of water flows at junctions. An analysis of this issue was investigated in considerable depth by Australian researchers operating for the Australian standards authority

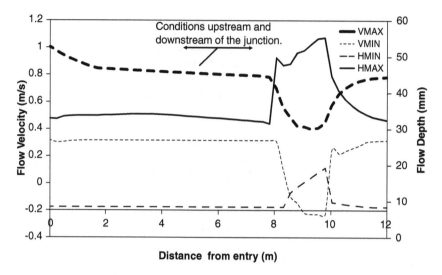

Figure 4.46 Flow velocity and water depth along the main drain line with discharge from top entry 45° junction.

(Cummings, 2010) working on the Australian/New Zealand plumbing code (AS/NZS 3500). The work focussed on simulating real-life problem installations and attempted to draw generalised conclusions from which code and standards could be adapted for these new problems being encountered as a result of climate change.

The experimental test rig adopted for the study is shown in Figure 4.47. The photograph shows 45° junctions; however top entry and 90° junctions were also tested.

In addition to the observation of solid transport distances and urinal performance indicators, the work produced graphic illustrations of the general form of junction interaction. They are reproduced here to illustrate some of the important conclusions derived from the DRAINET simulations discussed above.

Figure 4.48 illustrates the effect that a 90° top entry junction has on solids entering the main pipe from a branch. It can clearly be seen that solids do not only travel in the flow direction, but flow upstream of the junction. This has the effect of pushing solids out of the prevalent flow direction, with the attendant probability of blockage in the absence of contributing flow from other discharging appliances. This phenomenon is caused by the effective negative velocities as predicted for this junction and illustrated in Figure 4.43 and Figure 4.44.

The situation is improved by the introduction of a 45° junction as shown in Figure 4.49. In this case the effects of interaction are much smoother and the impediment to water flow is less pronounced. This could also be seen in Figure 4.45 and Figure 4.46 which show the DRAINET predictions of the

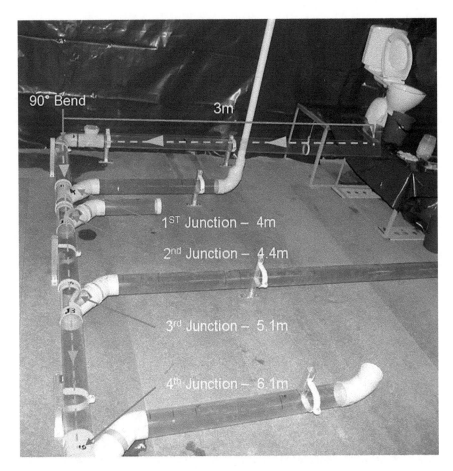

Figure 4.47 Experimental test rig used to test junction effects.

flow along the main collection drain line, in which the flow interactions were much smoother and less pronounced. Cummings' illustration from observations (Cummings 2010) shows that the solid transport characteristics of this junction are also improved.

The potential for blockages from level invert junctions is shown in Figure 4.50. The 'leakage' of water and waste material flow into the branch from the main drainline is caused by the upstream movement of the hydraulic jump. While this flow is likely to drain away again following the abatement of the hydraulic jump, it has the effect of reducing the force available to move solids once the event has ended. This can lead to an accumulation of waste and a higher risk of blockages. This phenomenon was also confirmed by Gormley (Gormley *et al.* 2013) in a large hospital complex drainage system.

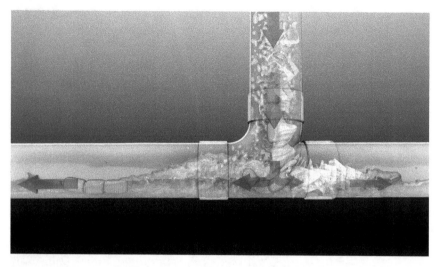

Figure 4.48 Effect of 90° top entry junction on solid transport.

Figure 4.49 A 45° top entry junction showing solids travelling with the
predominant flow.

And this is recorded from Cummings' laboratory investigations also, as
shown in Figure 4.51.

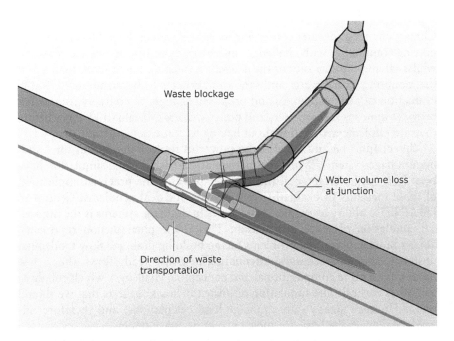

Figure 4.50 Potential for higher risk of blockages from level invert junctions.

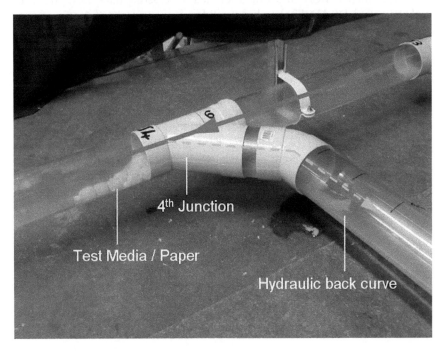

Figure 4.51 Laboratory confirmation of potential blockage risk from level invert junction. (Source: Cummings, 2010).

4.6 Concluding remarks

Climate change and water conserving measures present the public health engineering community with challenges hitherto unforeseen. Academic research, whilst salving the curiosity of the academic, must do much more than this if the mounting global water and sanitation crisis is to be surmounted. Rapid evaluation of new techniques and proposed changes to codes and standards provides a means for engineers and policy makers to evaluate changes before they are implemented, and without having to construct large test rigs.

This chapter has introduced the powerful tool of simulation for building drainage systems. While the mathematical basis for the simulation have been laid out in Chapter 3, this chapter presents the practical application of these methods to evaluating real designs and real problems. Central to DRAINET's ability to predict water flows in drainage systems is the numerical simulation of wave attenuation. This single phenomenon represents, almost in its entirety, the problems facing building drainage flow modelling: unsteady random discharges, attenuating to quasi-steady flows after a few metres, presenting computational and conceptual challenges which enlighten as well as enthral. The simulation of unsteady flows suggests that we should revisit Hunter's steady-state drainage load calculations, and therefore this work challenges the basis for most current drainage design.

No numerical model can be effective without validation through laboratory experimental work and validation against existing data. Both are dealt with here and DRAINET has been validated over nearly 30 years of concerted effort by many researchers. The validation of high velocity discharges and more tranquil Normal or Critical depth entry conditions, together with validation of rapidly varying flow and the moving hydraulic jumps associated with junctions, all make DRAINET a robust numerical model fit for this purpose.

Junctions have been found to be a potential source of increased risk of blockages in drainage systems subject to reduced water usage. At lower flow rates margins of error become important and the characteristics of some junctions over others present designers with choices that hitherto were not a concern. Blockages invariably involve waste material, and solid transport modelling is a major factor in performance assessment of a drainage system and is dealt with in detail in the next chapter.

5 Solid transport in building drainage networks

An essential requirement of building drainage networks is that waste solids are transported to a collection drain or sewer connection without undue deposition leading to maintenance issues and possible health hazards. It is inevitable that deposition will occur; however under satisfactory operating conditions such deposits are moved on by later appliance discharges or joining flows. It is essential therefore to develop a sound understanding of the parameters affecting and governing solid transport, deposition and subsequent motion. This requires a consideration of the properties of the network, in terms of the expected parameters of diameter, roughness, slope and length to the next flow confluence. In addition it is necessary to consider the frequency and type of appliance discharge to the network and the presence of any flow-enhancing devices. It is also essential to understand the mechanism of waste introduction to the network through the discharge profiles of any w.c. appliances served by the system, including the overall flush volume and most importantly, for the solid transport on the first flush, the solid position in the appliance discharge.

Waste solid transport in building drainage networks is dominated by the wave attenuation processes already discussed in Chapter 4. This provides a significant difference between solid transport considerations in building drainage networks, where the attenuation of the appliance discharge is important, and the quasi-steady flow conditions in larger diameter drainage and sewer flows that allow a more simplistic approach to the interaction between the applied flow and the solid motion. In particular wave attenuation introduces the concept of a maximum transport range for any appliance–network–solid combination: effectively the solid is deposited at the limit of the attenuated wave's ability to re-introduce motion.

There has been a continuous interest in defining waste solid transport in building drainage networks for at least the past 40 years. The research may be defined by two themes, namely an experimental theme undertaken across a wide range of international organisations and a much more restricted numerical modelling theme confined to a UK/US partnership and a single historic Japanese contribution. The experimental theme is sadly defined by an almost total lack of cross fertilisation due to a seeming lack of scientific method, particularly among commercial organisations undertaking testing, that has meant that excellent previous work has not been recognised and the wheel has been continuously re-invented – sometimes in triangular form. The simulation theme has been widely promulgated through publication in refereed journals and at conferences; however the lack of a literature search tradition or research method ethos has meant that it has not overly affected the practitioners of the experimental theme.

This chapter will present both themes and will show how the simulation of solid transport in building drainage systems can inform and extend the conclusions drawn from the experimental theme. The investigation of system frequency, junction position and shape and multi-storey discharge to a collection drain for example are not possible experimentally, and the simulation theme therefore has an important role in extending the understanding of transport issues.

5.1. Solid transport in building drainage networks – a historical perspective

The transport of deformable solids in the attenuating flows emanating from w.c. discharge has been a concern for a considerable time. Solid deposition leading to drain blockage may follow poor design and installation if the role of the various drain, appliance and solid parameters are not fully appreciated. Remedial action may be both costly and inconvenient, causing disruption to the building user.

The earliest recorded reference to solid transport as a criterion for either w.c. or system design may be traced to the Sanitary Institution, London, where in the 1890s it was demonstrated that a 9 litre flush was capable of clearing 95% of the test solids, in this case 12.5 mm balls, from the w.c. trap, depositing 21% along a 15 m long branch drain (Billington and Roberts 1982). This test is historically interesting as it predates by 100 years an almost identical test for 'drainline carry' developed for the US codes (ASME 2003). It is also interesting that this test was undertaken in order to persuade w.c. manufacturers to reduce flush volume, from the 36 litres common at that time, due to fears of a shortage of water in the metropolis – an almost direct precursor of discussions in the late 1990s that reduced UK flush volume to 6 litres (Defra 1999).

The next reported work on solid transport was undertaken at the National Bureau of Standards (NBS) in Washington, DC (Wyly 1964). Later Japanese researchers (Tsukagoshi and Matsuo 1975) developed suitable PVA model solids to allow a laboratory analysis of the mechanism of solid transport. Work at the Building Research Establishment in the UK (Lillywhite and Webster 1979) was also reported; however these contributions did not successfully define the mechanisms of solid transport, merely reporting distances travelled or per cent waste cleared in a series of test cases. Research at Brunel University from 1974 to 1980 concentrated on the mechanisms of solid transport and deposition, defining relationships between solid velocity and the surrounding flow conditions and has been fully reported (Swaffield and Wakelin 1976; Wakelin 1978; Swaffield 1980d). This work was undertaken initially in response to the waste solid deposition problems being encountered in a generation of new build hospitals that featured long, shallow-gradient drainage branches, often in inter-floor service voids; however it developed into a much more general research programme concerned with system operation under the influence of water conservation proposals then beginning to make themselves known in the UK following the drought of 1976. These relationships will be shown to be replicated by independent analysis following the development of suitable simulation techniques linked to a MoC simulation of wave attenuation and the introduction of a velocity decrement model to link solid transport velocities to the velocity of the surrounding water flow (Swaffield, 1980b, 1980c, 1980d).

The Brunel University experimental work up to 1980 was corroborated by substantial trials in an installed drainage network (Bokor 1982) utilising one of the new build hospitals mentioned. Bokor's work remains the only long term, 2 year, series of observations of faecal and waste transport in drainage branches served by a randomly used w.c. facility – in this case a hospital outpatients waiting area for both male and female patients and accompanying persons. This important contribution will be returned to later as its implications have not been fully understood; in particular Bokor's meticulous recording of faecal content of the w.c. flushes offers corroboration as to likely western diet faecal loads. In parallel Galowin (1979, 1982) at NBS investigated the forces acting on solids in transport, while Mahajan (1981a, 1981b) presented transport data for solids from laboratory work undertaken at the NBS facility, while other international contributions addressed w.c. discharge and solid transport (Neilsen 1973; Kamata, Matsuo and Tsukagoshi 1979). It will be seen that these references are to the annual meetings of the CIB W62 Working Commission on Water Supply and Drainage for Buildings that has met regularly since 1973, the 2011 meeting having been held in Portugal. This organisation has representation from across the international research community; however its papers are rarely referenced due to the lack of a literature search ethos among many of the organisations undertaking commercial testing to achieve short-term definitions of solid transport issues in drainage networks. This chapter will demonstrate that these contributions over 40 years hold many of the answers now sought as solid transport becomes important as water conservation criteria begin to dominate network design.

Thus an early impetus for this work was a need to reduce water consumption, and in particular water usage through w.c. flushing. In the UK this figure, prior to the introduction of a 6 litre flush in 1999, was generally accepted to be around 35% of the domestic potable water usage. Similar figures applied in most developed countries, with a probable higher figure in the US where flush volumes, prior to the 1992 introduction of a mandatory

6 litre flush, could be 36 litres. While these w.c. flush volumes had been accepted for some time, the need to economise had already been identified in Scandinavia where flush volumes as low as 3 litres, coupled to a system using 75 mm diameter drains at steep slopes, had been in use since the early 1970s; 'it is strange that the 75 mm diameter branch drains (sic) was a good w.c. drain only in Norway' (Rosrud 1977).

The report of the House of Commons Environment Committee (1996) entitled 'Water Conservation and Supply' identified the w.c. as the major cause of water consumption and recommended that 'reducing w.c. flush volume is a priority (that) could save 10% of water usage'. The committee felt that such savings would be a valuable contribution to offsetting the predicted effects of climate change. The committee also recommended the re-introduction of a dual flush w.c. operation to provide a lower flush volume to remove urine and tissue waste similar to the systems operating successfully in Australia. The committee's views were supported by water use data from the UK, US and Australia (Griggs and Shouler 1994; AWWA 1999; Cox 1997) that indicated that w.c. flushing in domestic applications absorbed from 23 to 32% of domestic potable water use, while in commercial applications some 40% was used for w.c. flushing, with a further 20% used in urinal flushing. Current estimates place UK w.c. domestic usage at around 25%.

Following the Environment Committee's report the Water Regulations Advisory Committee (WRAC) was established in 1996 and the resulting 1999 Water Regulations (Defra 1999) for England and Wales reduced w.c. flush volume to a maximum of 6 litres with a dual flush mechanism approved at a volume not to exceed two-thirds of the full flush – this flexibility allowing lower w.c. flush volumes to be considered.

The continuing concern as to water availability to support major UK housing projects implies that the concerns expressed by the Environment Committee remain relevant as evidenced by the recent formation of a Water Saving Group chaired by a UK government minister to deliver on the government's manifesto commitment 'to create a water saving body with the remit to cut water usage and encourage householders to use water more wisely' (Environmental Agency 2005). In the future the Building Regulations will include an emphasis on water conservation. The impact of climate change is recognised as a major issue, underlined by the UK Environment Secretary's comments in June 2009 (Benn 2009), on the release of the climate change scenarios for the UK from 2050 to 2080 (UKCP09) (Murphy *et al.* 2009) – 'climate change is the biggest challenge facing the world today. This landmark scientific evidence shows that we need to tackle the causes of climate change and deal with its consequences.'

While water conservation remains a prime objective in the design and provision of both domestic and commercial building infrastructures, it is also necessary to ensure that reduced water usage does not exacerbate other drainage system issues, such as satisfactory drain cleansing in the smaller bore drains and branches within the building envelope or within that part of the system that remains the responsibility of the building owner or user. While water conservation measures are aimed at an overall reduction in usage, new build proposals and changes in demography imply that over-all water usage as measured by sewer flows will not fall as steeply as the local flows within the building envelope or in the small bore connections to the sewer infrastructure. The governmental review of the Affordability and Supply of Housing (ODPM, now DCLG, 2001) and more recent demo-graphic analysis (DCLG 2009) makes it clear that demographic changes will have a major impact on the occupation norms for the UK. The review states that 'there will be a major increase in single person households which are projected to make up 39% of all households by 2026 compared to only 27% in 2001' (p. 7) – a increase of almost 50%. This implies that in-dwelling water use may decrease, but as the overall number of dwellings is set to increase by 4.8 million by 2026 compared to 2003, the sewer flows may not demonstrate a commensurate decrease. The review also identifies that a quarter of the single occupant households will be within the 55–64 age group with a second major contribution from single parent families.

However domestic water use is not based solely on per capita usage; data suggests that the overall dwelling usage is to some extent dependent upon the population housed in each unit. Edwards and Martin (1995) suggest that the effect of shared accommodation on the overall dwelling water consumption can lead to a 40% per capita usage for single occupants. This effect is impor-tant when the changing population demographic is included in the response to climate change predictions of water demand. Over the next three decades it is expected that a third of all new domestic dwelling in the UK will be single occupancy, thus adding to the likely water usage to be met under conditions of climatic change. Therefore water conservation and the identification of new or re-useable potential water resources must be allocated to offset water shortages in the domestic environment.

Thus the impact of appliance re-design to reduce water usage will have a disproportionate effect on conditions close to the building. It is in this region that flow conditions are also at their most complex as the appliance discharge undergoes wave attenuation, an effect characterised by a reduction in discharge peak instantaneous flow and depth as the wave traverses the initial branch drain. This reduction in peak flow conditions, accompanied by a 'spreading' of the wave is well known (McDougall and Swaffield 2000) and has the effect of reducing the transport capability of the flush wave. Further downstream the sewer flows tend to be quasi-steady so that the flow condi-tions are much simplified and do not require the application of unsteady free surface flow analysis. Satisfactory waste transport in the unsteady flow regime encountered in the upper reaches of the system therefore remains a central concern, particularly as w.c. flush volumes are progressively reduced and as the application of the 1999 Water Regulations leads to increased use of dual flush w.c.'s. These considerations are further exacerbated by changes in hygiene, in particular a movement towards increased usage of disposable products such as surface wipes that are often flushed.

The importance of both w.c. design and details of the network pipe diameters and cross-sectional areas was recognised and investigated during 1978 to 1982. Marriott (1978) and Swaffield and Marriot (1978) extended the Wakelin empirical model for solid transport and deposition to include drains of 75 mm diameter and parabolic cross section, the latter achieved by testing solid transport following w.c. discharge to an open 'deepflow' gutter. Reductions in acceptable flush volume were accomplished by the design and manufacture of a 6 litre flush volume w.c. (Uujamhan 1981), in conjunction with the UK Confederation of British Ceramic Sanitaryware Manufacturers (CBCSM), a w.c. installed and accepted on the Brunel University campus. Although CBCSM decided to set aside this advance, a follow-up design lowering the flush volume to 3 litres was successfully manufactured by Twyfords as part of a Brunel University/Overseas Development Administration (ODA) research programme that successfully installed the w.c. in Botswana, Lesotho, China and Brazil from 1984 to 1990 (Swaffield and Wakelin 1996; Bocarro 1987) The combination of low flush volume w.c. design and alternative drain cross sections as a solution to low flow solid transport was taken further by innovative work at Caroma, Sydney from 1996 onwards (Cummings 2003; Cummings, McDougall and Swaffield 2007).

By the end of the 1980s therefore there was a substantial body of experimental research that defined the mechanisms of solid transport, and the next phase was to be the introduction of mathematical modelling to allow the simulation of solid transport under a range of conditions and system parameters not readily provided by laboratory testing. As already discussed, the need for such an unsteady flow simulation had been recognised much earlier by Hunter (1940), Wyly (1964) and Burberry (1978). Thus from 1980 onwards the two identified themes, experimental and computational, overlap and interact. Experimental work has of course continued in isolation from the simulation capabilities developed due to the lack of a literature search ethos, as already indicated.

While experimental investigations defined the mechanisms of solid transport and demonstrated conclusively that reduced flush volume w.c. operation was inherently feasible, further progress required a means of simulating the attenuating flows present in the upper reaches of a building drainage system. A w.c. discharge to a drain will initially replicate the flow–time characteristic of the w.c., namely a peak flowrate of 1.5 to 2 litres/second with up to a 10 second duration depending upon the w.c. design, flush volume and mechanism utilized. The free surface unsteady flow encountered in building drainage systems is therefore characterised by the propagation of an attenuating wave such that the peak flow and depth downstream decreases while the flow duration increases. This results in a set of attenuating forces that act on the solid, leading to eventual deposition. A numerical simulation (Bridge 1984) of attenuating free surface flows based upon the method of characteristics to solve the St Venant equations (Lister 1960) allowed the prediction of the time-dependent free surface flow conditions throughout a building drainage network, as already discussed in preceding chapters. This development was part of the major Brunel University/NBS collaboration undertaken from 1980 to 1986, fully reported by Swaffield and Galowin (1992). Solid transport was included in this research effort from its inception; initially it was hoped to incorporate individual solid boundary conditions via the introduction of the forces acting on the solid and the presence of a moving hydraulic jump 'tethered' to the trailing edge of the solid; however this approach proved numerically tedious and was not developed further when a more convenient virtual solid velocity decrement model was introduced following consideration of the transport of 'floating solids' (Swaffield 1982, 1983; Swaffield and Galowin 1989).

An important determinant of the solid transport capability of any appliance and system was the recognition during the Brunel University/NBS collaboration (Swaffield and Galowin 1992) that there is an effective maximum range for any single solid discharged from a particular appliance to a drain and then subjected to repeated discharges from the same appliance (Swaffield and Galowin 1993). The solid will be deposited and moved on by each flush and then the next. However eventually the solid will deposit at a distance from the appliance where the discharge wave will have attenuated to the extent that it no longer has the 'energy' to re-initiate motion. This concept has now been generally accepted, Butler and Davies (2004) crediting its identification to Swaffield and Galowin, and is included as a determinant of system operation in recent solid transport assessments (Simms, Littlewood and Drinkwater 2006; Drinkwater, Moy and Poinel 2009). This concept will be returned to later.

The linking of the MoC model to a set of moving boundary conditions representing solid transport via a solid velocity decrement factor (McDougall 1995; Swaffield and McDougall 1996) allowed solid transport efficiency to be predicted for the first time, rather than inferred from experimental studies and site trials. This model was central to the discussion of proposals incorporated into Part H of the Building Regulations, ODPM (now DCLG) (2001) where McDougall's simulation (McDougall and Swaffield 2003) was successfully employed to ensure that the regulations did not exacerbate the probability of solid deposition by reducing the minimum diameter of the house drain to sewer connection in the reduced flows mandatory following publication of the Water Regulations (Defra 1999).

The application of McDougall's velocity decrement methodology, to be discussed in detail later in this chapter, allowed a range of investigations to be developed further at Heriot-Watt University, notably through a series of EPSRC-funded research projects covering such issues as defective drain interaction with solid transport (Swaffield, McDougall and Campbell 1999) and, through London Underground funding, a study of solid transport in the drainage networks of sub-surface installations (Wright 1997). The issue of the simulation of multiple solid transport interaction, an essential factor in waste removal, was initially investigated experimentally (Crerar 1989) and later in a combined simulation and experimental validation study (Gormley 2005; Gormley and Campbell 2006a, 2006b, 2007) that included the effects of adding surfactants in the form of household washing and cleansing agents to the flow.

Following the publication of McDougall's velocity decrement model (McDougall and Swaffield 2003; Swaffield and Galowin 1993), further experimental theme work continued to be undertaken to extend the understanding of the transport of gross solids in larger diameter sewers (Davies, Butler and Xu 1996; Littlewood 2000; Littlewood and Butler 2003), based on laboratory work utilizing the rigid cylinder model solids first developed by Galowin (1979). (Note these solids, hollow plastic cylinders of various diameters, are referred to in this text by their initial title of NBS solid that recognises their initial use, however this solid has been widely adopted and adapted since its use at Heriot-Watt University in the 1990s and is now sometimes found referred to in the literature as the Westminster solid, following its use by Butler and Davies (2004).) However the application to larger diameter sewers reduced the complexity (Davies, Schluter, Jeffries and Butler 2002) of the simulation necessary to determine the surrounding flow conditions as such sewer flows may be classified as quasi-steady and may be adequately described by use of commercially available packages (Butler, Davies, Jeffries and Schutze 2003) such as Hydroworks. Whether such a technique may be applied closer to the building is perhaps questionable, although the conclusions of such a study by Lauchlan, Griggs and Escarameia (2004) suggest that this is acceptable provided that the appliance discharges have attenuated sufficiently and combined to provide a near quasi-steady flow. Interest has continued unabated, for example the Walski, Edwards, Helfer and Whitlam (2009) study to determine the conditions for motion inception.

Extensive site monitoring (Bokor 1982) demonstrated that series interaction was central to the transport of waste solids following w.c. operation. In particular it was found that waste tissue following faecal material tended to dominate transport performance, a result now generally acknowledged (Drinkwater, Moy and Poinel 2009; Cummings 2010) and further demonstrated by the Gormley (2005) extension to the modelling of single solid transport to include series solid interaction by introducing a dependence on local wave propagation velocity to replace the velocity decrement relationship, a model that will be discussed later.

Thus the research community's response to the requirement that the mechanism of solid transport had to be fully understood has been substantial and has developed both a firmly based experimental understanding and a set of boundary conditions that may be used to determine both single and multiple solid transport in the attenuating flows found in building drainage systems and the less complex quasi-steady flows encountered in larger diameter sewers. The objective of this research programme, that has involved many national laboratories and academic institutions, was to aid in the reduction of water use by w.c. operation. This objective has been met as the operational flush volumes prescribed by national codes has fallen internationally, 6 litres in UK, USA, Europe, Australia, 4.5 litres in Singapore, 3 litres allowable in Scandinavia etc. Coupled to this has been the successful re-introduction of dual flush, with 6/3 litres being generally accepted, while 4/2 litre dual flush w.c.'s are available.

Paradoxically the effort may have been too successful as there are now concerns, originating from groups based around appliance manufacturers, that water conservation may lead to such a reduction in throughflow that deposition will become a major concern. These concerns found an outlet at the 'Dry Drains Forum' held in Frankfurt in 2009 and in presentations to the CIB W62 2009 conference in Dusseldorf, and in the initial objectives of the Plumbing Efficiency Research Coalition (PERC) formed in the US with representations from manufacturers and code bodies. PERC aspires to undertaking new research into drainline carry that would involve university CFD modelling, test laboratory validation and field studies by plumbing contractors. Funding has to date been elusive, but a collaboration with the Australasian Scientific Review of Reduction of Flows on Plumbing and Drainage Systems (ASFlow) activity is promising (de Marco 2009; Cummings and de Marco 2010).

Similar concerns may be read into reports by WRc in the UK (Drinkwater, Moy and Poinel 2009). Whether such concerns are justified is questionable when it is considered that much of the drainage infrastructure has been in place for decades and has seen a very substantial growth in water usage. Hence a reduction to levels common 10 to 15 years ago should not necessarily cause a problem. Swaffield (2009a) presents an alternative view, based on existing research that covers almost all the research requirements identified by PERC (de Marco 2009), that encourages a review of the existing published knowledge from almost 40 years of research prior to embarking on a further reinvention of the wheel; the fact that none of the material cited was generally known to PERC[1] re-emphasises the need for a literature search prior to testing ethos.

Thus the experimental theme has flourished for 40 years and is continuing regardless of the body of work available for review, as evidenced by Walski (2011). The simulation and analysis theme remained a major research activity at Heriot-Watt University based around the MoC finite difference solutions. Initial work on CFD representations of solid transport (Oengoeren and Meier 2007) has a long development period ahead before it can match the speed and applicability of the proven one-dimensional MoC solutions. Solutions based on commercially available drainage packages, such as Hydroworks, are limited to quasi-steady flows and cannot contribute to the assessment of solid transport in an attenuating flow regime. The development of the initial Brunel University/NBS research program into the DRAINET simulation already discussed and capable of delivering solid transport predictions and the development of alternative models to McDougall's to allow the inclusion of multiple solids and a representation of the flow depth variations across the solid, omitted from the virtual solid decrement simulation, are the only current activity in this theme. It is essential that research effort is given to developing alternative DRAINET models as a monopoly is never a sustainable position. The development of alternatives to the MoC solution to the St Venant equations is a possible line of future work, building possibly on the Lax-Wendorf solution once suggested by Kamata, Matsuo and Tsukagoshi (1979) but never developed.

This chapter will analyse two substantial data sets that demonstrate clearly the effect of w.c. flush volume on the transport of deformable solids. These data sets allow the influence of the system parameters of branch drain slope and cross-sectional dimension and, uniquely, shape to be investigated. Reducing flush volume with no regard to drain diameter or slope will inevitably lead to an increased probability of solid deposition. In addition it has been clearly demonstrated (Swaffield and Galowin 1992) that in any system there is a maximum transport distance for any combination of solid/branch drain parameters subjected to repeated w.c. flushing. Effectively the solid is finally deposited at a transport distance commensurate with the attenuation of the applied w.c. discharge to the point where it no longer has the depth or specific energy required to initiate further solid movement. Thus a stance that depends on subsequent w.c. flushing or other appliance discharge to enhance transport in a building branch drain is flawed.

Reducing branch drain diameter in step with w.c. discharge is an attractive proposition and is now allowed in European codes. However there are limitations due to the range of waste materials introduced to the drainage network by w.c. operation. At the current time a 75 mm diameter branch drain to serve low flush w.c. discharge appears a reasonable lower limit – as originally proposed in Scandinavia. However there is an alternative which is to consider non-circular cross sections. Elliptical or parabolic cross section drains may be a suitable alternative for exactly the same hydraulic reasons that led Victorian engineers to use non-circular sections in the major sewer constructions of the nineteenth century. At low flows both the flow depth and velocity are enhanced compared to a circular section which should lead to enhanced solid transport based on a velocity decrement model. However if the solids in transport are deformable and large then increased wall friction may be a negative consequence of reduced channel cross section at low flows.

Two major solid velocity in transport data sets are available for re-analysis. The first set is laboratory based and was initially reported by Swaffield and Marriott (1978) as a natural extension to the work reported by Wakelin. This data set includes two w.c. types, characterized by trap seal volume, four flush volume settings, from 9.1 litres down to 4.5 litres, three drain cross sections, including a 100 mm and 75 mm diameter pipe and a parabolic open channel, as well as seven branch drain slopes; a data set of some 100 test cases, each based on up to 200 solid velocity measurements. The parabolic channel had a top width of 75 mm at 108 mm depth.

The second set is the outcome of the ASPM Flushability Test (Howarth, Swaffield and Wakelin 1980), established to provide a basis for a declaration of 'flushable' on a series of female hygiene products that required 94/100 product samples sequentially flushed from a BS w.c. to travel at least 10 m along a 100 mm UPVC drain set at 1/80 – parameter choices commensurate with UK domestic drainage. (The choice of 10 m was made on the basis that this would cover the majority of drain lengths in the UK between the road-way sewer and the bathrooms installed in typical UK urban strip housing).

The re-analysis of both these data sets will allow a consideration of the effect on system performance of w.c. flush volume reduction as well as indi-cating the possible advantages of reduced cross section drains.

5.2 Mechanism of deformable solid transport in attenuating flows

The flow conditions at the head of a branch drain and subsequently down-stream are governed, as already discussed, by the attenuation of the appli-ance discharge. The leading edge of the discharge wave tends to steepen as the later, deeper flow overtakes the initial discharge. Conversely the tail of the wave spreads as the deeper flow outdistances the trailing flow. Consequently the prediction of wave attenuation and its effect on solid transport in build-ing drainage networks is a substantially more complex issue than the predic-tion of solid transport in the quasi-steady flows to be found in the larger diameter drains downstream.

Figure 5.1 illustrates the zonal concept of solid transport (Wakelin 1977). On discharge from the w.c. the deformable solid impacts the drain invert and is compressed further by the force of the increasing flow discharged behind the solid as the appliance reaches its peak discharge rate. The solid decel-erates markedly; however this deceleration allows following water to dam up behind the solid, thus generating an accelerating force that is strength-ened by the momentum transfer from the increasing discharge. The solid re-accelerates and may reach a velocity in excess of its initial discharge value; Zone 1, Figure 5.1, illustrates this effect.

The deformable solid is then carried downstream and is subject to a sys-tem of forces that includes momentum transfer from the surrounding water flow, buoyancy due to flow depth, gravitational and frictional forces, the latter due to contact with the channel surfaces. In deeper water the solid tends to re-adjust to its natural form, the compression observable on w.c. discharge being replaced by an extension due to the shear forces exerted by the surrounding flow and the effects of buoyancy. This Zone 2 transport is characterized by a minimum depth differential across the solid and a solid velocity that may be described by a decrement factor applied to the sur-rounding flow (McDougall 1995).

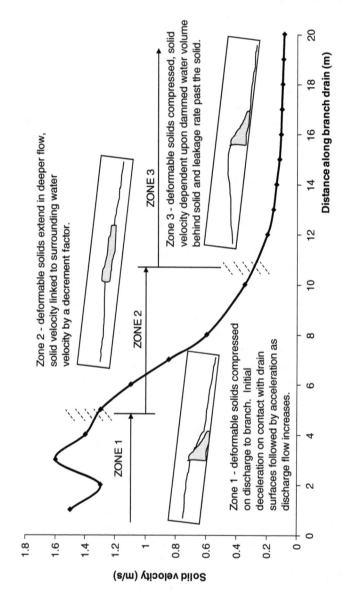

Figure 5.1 Zonal description of the mechanism of solid transport in attenuating flows in a branch drain following a w.c. discharge.

However, as transport continues the solid decelerates as the accompanying wave attenuates, establishing a decrement between solid and water velocity. The overall impact of the decrement factor is to move the solid back relatively through the accompanying flush until the surrounding flow is insufficient to support Zone 2 transport. At this stage the solid tends to 'ground' and the frictional forces predominate. This deceleration results in a damming up of flow behind the solid and the establishment of a hydrostatic force–based transport regime – Zone 3. As the w.c. discharge is finite, the volume of water available to dam behind the solid is limited so the duration of the hydrostatic driving force maintaining solid motion is determined by the leakage flow across the solid. Zone 3 is thus characterized (Figure 5.1) by a high depth differential and a low leakage across the solid. The solid velocity is low, and the duration of transport depends on the relationship between depth differential and leakage. Any disruption to this slowly changing flow condition, such as passage through a junction or over a badly made pipe joint, may allow enhanced leakage, a diminution of the available dammed water and lead to deposition. Deposition in Zone 3 may take place at any velocity below 0.2 m/s.

While Figure 5.1 illustrates the zonal mechanism of solid transport, Wakelin (1977) also developed a general relationship that allowed solid transport velocity at any gradient to be plotted on a common axis. Experimental investigation of the transport of a standard deformable solid along drains of a single diameter set at a range of gradients and subjected to a standardized w.c. discharge indicated a relationship:

$$V_{solid} = C_1 - C_2 \sqrt{\frac{L}{G}} \qquad (5.1)$$

so that the deposition position, $L_{deposition}$, at any drain slope, G, could be determined by setting solid velocity to zero:

$$L_{deposition} = G \left\{ \frac{C_1}{C_2} \right\}^2 = GC_3^2 \qquad (5.2)$$

Note that G is used to represent slope where reference is made to Wakelin's original model.

Figure 5.2 illustrates the $\sqrt{(L/G)}$ data presentation for a range of w.c. discharge volumes to a 100 mm diameter drain set at gradients from 1/40 to 1/300. The Zone 1 and 2 conditions are easily identified and suggest a deposition value of $\sqrt{(L/G)}$ from 16 at 4.5 litres flush volume to 37 at 9.1 litres. The Zone 3 low velocity transport region is also clearly identifiable for flush volumes of 7.5 litres and below, being particularly marked at 4.5 litres.

An interesting anomaly at 1/40 is also identifiable. The deformable solid discharged to the drain at this relatively steep slope is initially retarded, but

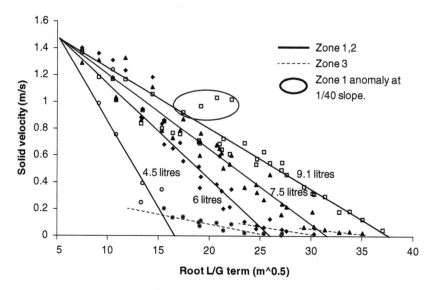

Figure 5.2 Deformable solid velocities measured in a 100 mm diameter branch
drain at a range of gradients and w.c. flush volumes.

the secondary acceleration phase is greatly assisted by the drain slope so that
the solid does not display the expected deceleration within the 14.4 m long
test drain available. It will be seen that at flatter gradients, or lesser flush
volumes, this effect is not recognizable.

Thus the extended test series (Swaffield and Marriott 1978) confirmed
Wakelin's hypothesis that the $\sqrt{(L/G)}$ term was central to an understanding of the mechanism of solid transport in attenuating flows. However to
extend this format to include variable flush volume and drain cross-sectional
shape requires a re-analysis to identify the fundamental parameters governing solid transport.

In order to analyse the data available on deformable solid transport in the
attenuating flows following various w.c. discharges to the head of a branch
drain of a particular cross-sectional shape and slope it is necessary to consider the parameters that affect solid transport.

The drain parameters are critical, namely drain slope, S, diameter, D, or
hydraulic mean depth, m, and wall roughness, k, together with solid cross-
sectional area, $w^{*}t$, length, l, and dry and saturated masses, m_{dry} and m_{wet}.

The w.c. discharge profile, i.e. duration, peak flowrate and overall volume, $Vol_{w.c.}$, determine performance, together with the actual water volume
discharged behind the solid, Vol_{Behind}. This latter volume is critical as it
provides the initial impetus to ensure transport and the residual water volume available to drive the solid beyond the Zone 2/Zone 3 interface. This
parameter is linked to the design of the w.c. trap, a major determinant of
efficient w.c. design (Uujamhan 1981).

In addition, reducing the flush volume of a w.c. designed to flush at a particular volume introduces a parameter Vol_{Actual} that may be less than $Vol_{w.c.}$. Reducing a w.c. discharge by absorbing cistern volume or inducing a break in the discharge at an intermediate water level tends to result in similar discharge profiles up to the point of interruption. The use of a term describing the ratio of actual volume to design volume is therefore a suitable shorthand for this effect.

The data set analysed included two w.c. types, namely a P trap 9.1 litre w.c., trap seal volume 1.89 litres, the 'BS' unit, and a horizontal outlet pan, trap seal volume 1.53 litres, the 'EU' unit. The discharge volume was reduced to 7.5, 6 and 4.5 litres by interrupting the siphon action as soon as the cistern water level fell to a pre-determined level consistent with the required flush volume by the simple expedient of introducing a flexible tube into the siphon that allowed air ingress as the water level fell.

Three drain configurations were considered, namely 100 mm and 75 mm diameter pipes and an open gutter of parabolic cross section, top width of 75 mm at a depth of 108 mm. In each case the channel material was UPVC so that the surface roughness, k, could be considered constant. Drain slope was varied from 1/40 to 1/200, and the solid velocity was recorded at eight locations along the 14.4 m drain length by means of pairs of photoelectric cells and light sources, interruption to the light emitted or reflected by the passage of the solid initiating a recordable time signal.

A single deformable sanitary product was used for all the tests analysed, length 300 mm, width 60 mm and thickness 20 mm. The wet and dry mass was 250 and 16.5 gms respectively. This solid deformed in the flow direction as a result of the forces generated by the surrounding flow and in cross section due to the form of the enclosing channel cross section. This product, referred to as a 'maternity pad', was one of the items whose transport was identified as problematic in the hospital surveys that led to Wakelin's initial work.

Figure 5.3 illustrates the experimental test rig used to determine the solid transport characteristics of deformable solids under conditions of reduced flush volume and drain cross section. The solid velocity measurement system was based on nine sets of variable photocell output as a light source was interrupted – in the case of transparent pipes – by the solid passing over the light source mounted on the pipe invert, or for the opaque pipes by a reflective system relying on the increased reflection from the white solid compared to the accompanying water. In the latter case both light source and photocell were mounted on the crown of the pipe – a suitable cutaway being provided in the case of the 75 mm diameter opaque pipe.

Flush volume ahead of the solid was measured in a collection tank by means of a water level sensing linear displacement transducer. Following each flush the collection tank was pumped down to a pre-set level in readiness for the next flush.

Figure 5.4 illustrates the three channel cross sections and the likely cross section of the solid. It will be seen that in the case of the parabolic gutter the

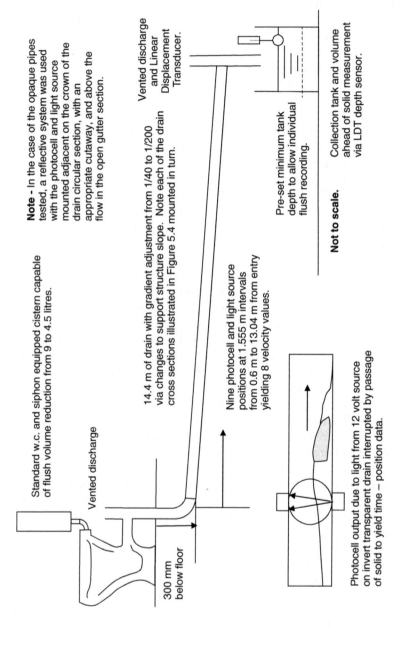

Standard w.c. and siphon equipped cistern capable of flush volume reduction from 9 to 4.5 litres.

Vented discharge

300 mm below floor

14.4 m of drain with gradient adjustment from 1/40 to 1/200 via changes to support structure slope. Note each of the drain cross sections illustrated in Figure 5.4 mounted in turn.

Nine photocell and light source positions at 1.555 m intervals from 0.6 m to 13.04 m from entry yielding 8 velocity values.

Photocell output due to light from 12 volt source on invert transparent drain interrupted by passage of solid to yield time – position data.

Note - In the case of the opaque pipes tested, a reflective system was used with the photocell and light source mounted adjacent on the crown of the drain circular section, with an appropriate cutaway, and above the flow in the open gutter section.

Vented discharge and Linear Displacement Transducer.

Pre-set minimum tank depth to allow individual flush recording.

Not to scale.

Collection tank and volume ahead of solid measurement via LDT depth sensor.

Figure 5.3 Experimental variable slope and cross section test rig to determine the solid transport characteristics under reduced flush volume conditions.

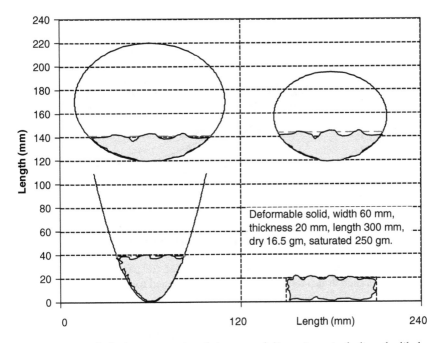

Figure 5.4 Branch drain cross-sectional shapes and dimensions, including the likely cross-sectional shape of the solid during transport.

solid takes up to 40 mm of channel depth while in the 100 mm and 75 mm diameter drains the solid absorbs cross-sectional area up to 21 and 24 mm depth respectively.

Figure 5.5 illustrates typical solid transport performance following the discharge of a 6 litre flush volume into each branch drain set at a slope of 1/60. The four test cases are presented in ascending order of solid velocity along the drain. The 'best' performance, namely the highest solid velocity over the length of the drain, occurred in the 75 mm diameter branch, the worst occurred in the 100 mm diameter drain. The larger trap volume w.c. generated the worst overall performance. These results were expected due to wave attenuation in the branch drain. Figure 5.6 illustrates the mechanism of wave attenuation following a 'water only' flush to a branch drain, confirming the reduction in peak flow depth as the wave propagates downstream, accompanied by a reduction in peak flow velocity and hence reduced forces available to act on the solid.

As surrounding flow depth is a central factor in solid transport and while the depth in the open gutter is the greatest, it might be expected that the parabolic section would result in the higher solid velocities, an expectation not confirmed by Figure 5.5. The contact perimeter between the deformable

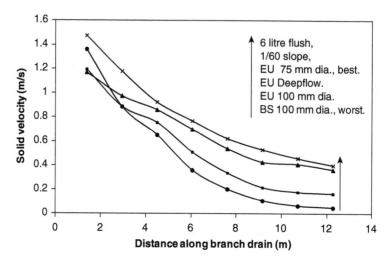

Figure 5.5 Solid velocities recorded along the circular and parabolic cross section branch drains tested at a 1/60 slope subjected to a 6 litre flush.

solid and the channel walls in the parabolic section is also the greatest, so frictional forces may negate the buoyancy advantages of increased depth. Analysis of the full data set will clarify these issues. Figure 5.6 also presents the flow maximum specific energy values following the discharge of a w.c. flush to the simulated drain. Specific Energy, E, may be defined as:

$$E = h + \frac{u^2}{2g} \tag{5.3}$$

where u is the mean flow velocity and h the local flow depth. Again the parabolic gutter has the highest free surface flow specific energy at all locations; however the solid transport performance will inevitably depend on a combination of factors, including local flow velocity and depth and the frictional forces dependent upon surface contact area or wetted perimeter.

The solid velocity may be considered to depend upon the following parameters:

$$V_{solid} = \varphi(Vol_{w.c.}, Vol_{Behind}, Vol_{Actual}, S, L, m, k, solid$$
$$- parameters, g) \tag{5.4}$$

As the solid parameters and wall roughness are constant:

$$V_{solid} = \varphi(Vol_{w.c.}, Vol_{Behind}, Vol_{Actual}, S, L, m, g) \tag{5.5}$$

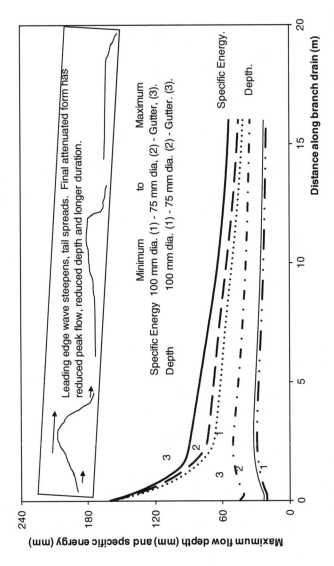

Figure 5.6 Predicted peak depths and specific energy values along the various branch drains considered as an example of the effect of wave attenuation.

From Buckingham's π Theorem, as these parameters are in two dimensions only, length and time, and as there are eight variables, the dimensionless expression will include six groups where m and g would be suitable repeating variables.

The dimensionless form of the expression thus becomes

$$\frac{V_{solid}}{\sqrt{mg}} = \varphi_1\left(\frac{Vol_{w.c.}}{Vol_{Actual}}, \frac{Vol_{Actual}}{Vol_{Behind}}, \frac{Vol_{w.c.}}{m^3}, S, \frac{L}{m}\right) \tag{5.6}$$

$$V_{solid} = \sqrt{mg}\varphi_1\left(\frac{Vol_{w.c.}}{Vol_{Actual}}, \frac{Vol_{Actual}}{Vol_{Behind}}, \frac{Vol_{w.c.}}{m^3}, S, \frac{L}{m}\right) \tag{5.7}$$

Note that the group $\dfrac{V_{solid}}{\sqrt{mg}}$ is a form of Froude Number linking solid and water velocity and depth, a concept picked up later by Gormley (2005) and Walski *et al.* (2011).

Combining terms to recognize the importance of the $\sqrt{(L/G)}$ term and introducing indices that may be determined by regression analysis, Table 5.1, yields:

$$V_{solid} = \sqrt{mg}\varphi_2\left\{\sqrt{\frac{L}{mS}}, \left\{\frac{Vol_{w.c.}}{Vol_{Actual}}\right\}^\alpha, \left\{\frac{Vol_{Actual}}{Vol_{Behind}}\right\}^\beta, \left\{\frac{Vol_{w.c.}}{m^3}\right\}^\gamma\right\}$$
$$= \text{Function} \tag{5.8}$$

For each of the systems investigated, data exists to define the solid velocity at 8 positions along a 14.4 m branch drain. Figure 5.5 has already demonstrated the scale of the velocities, typically in excess of 1.4 m/s on drain entry reducing to values below 0.4 m/s downstream. Solid deposition position was also recorded in those cases, usually at shallow slope or reduced flush volume, where the solids did not clear the test drain. Previous research identified the benefit of dimensional analysis by introducing the $\sqrt{(L/G)}$ term; however in this case there is a need to refer to equation 5.8 to introduce the effects of drain cross section and reduced flush volume. Figure 5.7 demonstrates the application of equation 5.8 to organize the data set applicable to one w.c. type/drain cross section combination for a range of flush volumes. It is clear that the zonal model developed by Wakelin applies. The central Zone 2 dominated by the velocity decrement approach to solid velocity is clearly identifiable. Similarly Zone 3 transport is preserved, reinforcing the role of dammed water behind the solid in generating hydrostatic forces sufficient to continue solid motion. It is clear that a linear relationship linking solid velocity to the equation 5.8 function exists with the appropriate indices listed in Table 5.1.

Figure 5.8 extends this approach to the whole data set by plotting mean solid velocity at each of the eight monitoring points for each of the test cases

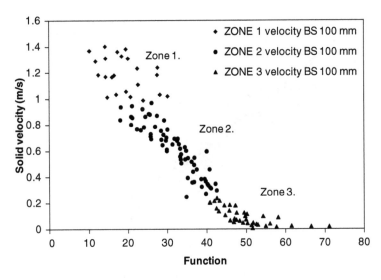

Figure 5.7 Solid velocities for all branch drain cross sections, BS w.c. flush volumes and slopes, demonstrating the result of the regression analysis for one test case.

Table 5.1 Dimensionless group indices (equation 5.8)

W.c. type	Trap volume, litres	Average volume, litres	Drain type	Hydraulic mean depth, mm	α	β	γ
BS	1.89	1.84	100 mm dia.	D/4 = 25	0.36	1.36	0.179
EU	1.53	1.43	100 mm dia.	D/4 = 25	0.36	1.55	0.179
EU	1.53	1.43	75 mm dia.	D/4 = 18.75	0.36	0.9	0.159
EU	1.53	1.43	Open gutter width 75 mm at 108 mm depth	A/P = 22.9	0.40	1.1	0.173

against the local value of the equation 5.8 function, effectively the $\sqrt{(L/G)}$ term modified by factors dependent on the system parameters. The general form of equation 5.8 allows inferences to be drawn as to the necessary drain conditions to accommodate w.c. flush volumes or to guarantee minimum solid transport distances.

In order to determine the form of equation 5.8 for each w.c. type it was necessary to record the water volume discharged behind the solid in each case. Table 5.2 presents the measured values for each of the two w.c. types considered and the trap seal volume for each. The solid itself will affect the volume discharged prior to and behind it from the w.c. In the data set analysed the deformable solid provided a substantial blockage in the w.c. trap

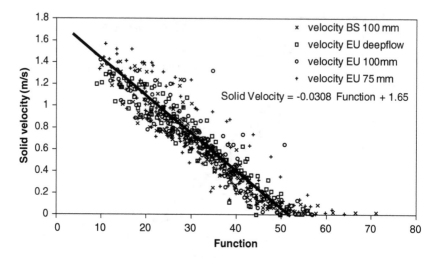

Figure 5.8 Solid velocities for all branch drain cross sections, w.c. type, flush volumes and drain slopes, demonstrating the result of the regression analysis for all the cases considered.

Table 5.2 Volume discharged ahead of solid at each reduced flush volume setting

W.c. type	EU	EU	BS	BS
Flush volume, litres	% flush ahead of solid on discharge	Volume ahead of solid on discharge, litres	% flush ahead of solid on discharge	Volume ahead of solid on discharge, litres
9.1	16.70	1.51	17.80	1.60
7.5	20.20	1.51	27.10	1.6
6.0	24.60	1.48	31.30	1.88
4.5	27.50	1.24	40.80	1.84
Mean volume ahead of solid, litres		1.43		1.84
Trap seal volume, litres		1.53		1.89

and therefore the solid left the w.c. roughly following the discharge of the trap seal water. It will be appreciated that increasing the water volume discharged behind any solid will improve transport performance, as will increased total flush volume, reduced cross section and increased slope – all parameters whose influence may be inferred by reference to equation 5.8 and Figure 5.7.

While the general form of equation 5.8 is based on velocity data, and while the transport limit of the function is set at 53.5 (Figure 5.8), the data set also includes deposition positions for those cases where, predominantly at shallow gradients and low flush volumes, deposition occurred along the length of the drain. Figure 5.9 demonstrates generally good agreement between observed solid deposition and the transport distances determined from Figure 5.8. There is a tendency to underestimate travel distance at the lower flush volumes and shallower slopes. Agreement improves as these parameters increase. Overall the re-analysis presented allows observations as to the effect of reduced flush volume on transport performance.

As demonstrated by the form of equation 5.8, and by Figure 5.8, transport is enhanced by reducing the volume of flush discharged ahead of the solid. Similarly increasing overall flush volume and branch drain slope while reducing drain cross section improves performance. Thus detailed w.c. design is essential to achieving acceptable solid transport at reduced flush volumes. The w.c. types considered in this data set were typical of the majority of the then installed appliances; however efforts to improve design will have an increasing impact as the natural replacement programme continues. Reducing drain cross section is a major opportunity as w.c. flush volumes and other water conservation measures tend to reduce domestic and other building flows. It is therefore essential that regulations governing drain sizing take into full consideration the effects of reduced water flows on solid transport.

A natural development of reducing drain cross section is to consider non-circular sections. Sewer experience suggests that non-circular sections may have advantages at low flows as the flow depth, and hence the buoyancy of any solid, is enhanced leading to improved transport efficiency. In the data set considered this effect was probably masked by the choice of a relatively large deformable solid so that, despite the flow depths being increased, the

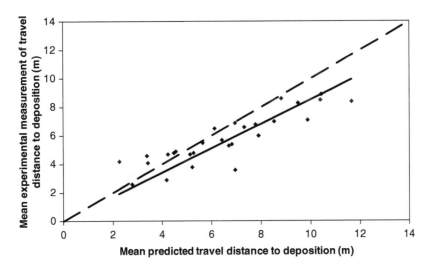

Figure 5.9 Solid deposition data compared to the predicted mean travel distance to deposition indicated by Figure 5.8.

greater surface contact probably led to the rank order where a 75 mm diameter drain performed slightly better than the parabolic gutter. The original analysis concluded that there was scope for more research in this area, a suggestion taken up by Cummings (2003).

Figure 5.10 illustrates the application of equation 5.8 to yield design drain slopes for two required transport distances, 5 and 20 m, for a range of flush volumes from 9.1 litres, then still in substantial use in the UK, through the installed w.c. stock, down to 4.5 litres. It is stressed that the choice of acceptable transport distance depends entirely on the system configuration. The basic premise is that a single isolated w.c. should be able to ensure transport of waste solids to the first junction with joining flows from other appliances or buildings. It is clear that the minimum branch drain slope required increases as flush volume decreases for all cases.

The increase in the usage of flushable sanitary products in the late 1970s led the Association of Sanitary Protection Manufacturers (ASPM) to develop an industry standard for flushability which remained in use for 30 years (Howarth, Swaffield and Wakelin 1980). In order to ensure statistically valid results the standard defines a 100 flush test series. A sanitary product placed in a BS w.c. must leave the w.c. on the first flush and travel 10 m along a 100 mm diameter branch drain, at a slope of 1/80, 94 times out of 100. A test sample failing to leave the w.c. on the first flush must be discharged on a second flush to avoid failure. This test introduced for the first time the concept of both appliance and drainline carry performance testing.

The extensive testing of commercially available sanitary products provided a source of data to corroborate the transport model, the second data set to

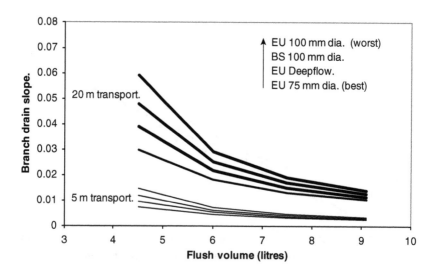

Figure 5.10 Branch drain slope to achieve a particular solid transport performance, indicating the steepening necessary as flush volume is decreased.

be re-analysed here. This standard, which continued in use for 30 years, was based on the need to ensure both w.c. discharge and subsequent drain transport for a distance sufficient to reach junctions with other appliance flows: a minimum transport distance of 10 m in a 100 mm diameter drain at a slope of 1/80 following discharge from a 9.1 litre BS w.c. being determined to be representative of the typical strip housing in the UK. The second data set is therefore drawn from a wide range of sanitary products subjected to this assessment: product length, width, thickness, wet and dry masses and V_{Solid} vs $\sqrt{L/G}$ results were all recorded making this data set a unique collection of test solid transport performance data.

Recent developments in data analysis support have allowed this body of data to be re-assessed to yield guidance on the implications for solid transport of variations in both system and solid parameters: that analysis is presented here for the first time. Observation suggests that Wakelin's $C_{1,2}$ coefficients were dependent upon product dimensions of length, width and thickness, l, w, t, and mass, m, wet and dry, as well as the branch diameter, d, and gradient, S. Water density, ρ_w, the actual flush volume, F, and the volume discharged behind the solid, F_B, are also relevant. Reducing flush volume, analogous to a lower dual flush capability, introduces the w.c. design flush volume, F_{wc} as a variable. Gravitational acceleration g is required as the flow is free surface. The transport coefficients $C_{1,2}$, may thus be expected to be functions of

$$C_{1or2} = \phi(l, w, t, m_{dry}, m_{saturated}, D, S, F, F_B, F_{w.c.}, g, \rho_{water}) \tag{5.9}$$

Buckingham's π Theorem suggests that, as these variables display dimensions of Mass, Length and Time, the 13 variables may be transformed into 10 dimensionless groups. The repeating variables are ρ_{water}, g, and D.

$$\pi_1 = F/F_B, \qquad\qquad \pi_2 = l/D, \qquad \pi_3 = wt/D^2,$$

$$\pi_4 = t/D, \qquad\qquad \pi_5 = F/D^3, \qquad \pi_6 = m_{saturated}/m_{dry},$$

$$\pi_7 = m_{saturated}/(\rho_{water}ltw) = sg, \qquad \pi_8 = F/F_{wc}, \qquad \pi_9 = S,$$

$$\pi_{10} = C_1/(\sqrt{g}\sqrt[6]{F_B}) \text{ or } \pi_{10} = C_2/\sqrt{g}$$

$$\pi_{10} = \frac{C_1}{\sqrt{g}\sqrt[6]{F_B}} = 0.755 + 0.00205\left(\frac{\pi_1^{0.622}\pi_4^{2.866}\pi_5^{0.886}\pi_6^{1.084}}{\pi_2^{0.677}\pi_3^{2.297}\pi_7^{1.62}\pi_8^{2.53}}\right) \tag{5.10}$$

$$\pi_{10} = \frac{C_2}{\sqrt{g}} = 0.00442 + 0.0377\left(\frac{\pi_1^{0.74}\pi_2^{0.996}\pi_3^{0.051}\pi_4^{1.013}}{\pi_5^{0.616}\pi_7^{0.55}\pi_8^{0.69}}\right) \tag{5.11}$$

$$C_3 = \frac{C_1}{C_2} \tag{5.12}$$

As the results analysed were all at a single gradient, the slope term S is omitted.

Figures 5.11 and 5.12 illustrate the form of these dependencies for a range of commercially available sanitary products with width and thickness parameters from 45 to 70 mm and 3 to 20 mm respectively and lengths from 133 to 300 mm. Dry and saturated weights varied from 2 to 19 gms and 22 to 250 gms. Data for flush volumes from 9 to 4.5 litres was available.

Figure 5.12 illustrates the relative predicted performance in terms of the likely transport distance, the C_3 parameter, as the various parameter values are changed relative to a chosen datum, for example as the specific gravity group, π_7, is increased relative to a chosen datum of 0.2 or a cross section blockage, π_3, is increased relative to a datum of 5%. For all other parameters constant, increasing solid specific gravity, solid cross-sectional area or saturated mass all result in a reduced transport at any flush volume relative to the datum C_3. The design of the w.c. has a fundamental effect on solid transport as a poor w.c. design will discharge solids late in the flush resulting in a severely curtailed transport. Reducing w.c. flush volume reduces transport if the volume is decreased with no change to the w.c. design, significant for dual flush installations and retro-fit flush volume reducers from the 'urban myth' brick to drop valve installation discharge dams.

The re-analysis of these data sets confirms that the outcome of the long-term research effort, involving many national building research establishments and academic institutions, has been a better understanding of solid transport in the attenuating flows encountered in building drainage systems. This effort, based initially on experimental observation of solid transport

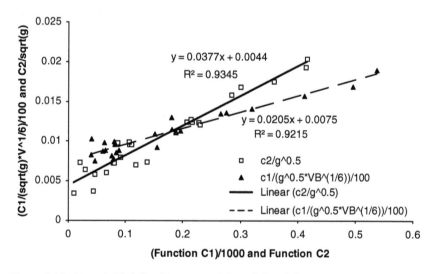

Figure 5.11 C1 and C2 defined in terms of the solid and drain parameters as identified in equations 5.10 and 5.11.

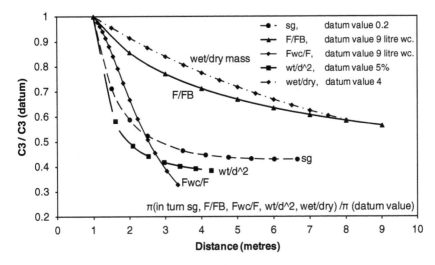

Figure 5.12 Dependency of deformable solid transport on a range of solid, appliance discharge and branch drain dimensionless groups.

in both the laboratory and in situ, will be shown to have contributed to the simulation theme mentioned earlier: the development of system simulations will aid the system designer as well as the w.c. manufacturer. Reducing flush volume remains the central plank of any policy to reduce domestic and commercial potable water usage. The re-analysis of the major data sets presented in this chapter has demonstrated the inherent linkages that exist between transport performance and both system and appliance design parameters and suggests that current proposals to reduce flush volume to 4 litres and below are feasible.

However the re-analysis has also shown that empirically derived expressions can only be applied within the bounds of the parameters included in the test programme. It is necessary to take the analysis further to allow predictions rather than extrapolations to be made with confidence. For this to be a believable development it is necessary to show that the empirical understanding accurately represents 'real' in situ system operation and that there is a reverse link from the simulations developed to both laboratory and site observations. Both of these criteria are met by the simulations to be introduced.

Figure 5.13 illustrates, by a re-plotting of data, the degree of fit achieved when the solid velocity data was recast in the form suggested by equation 5.1. Figure 5.13 also includes a minor improvement to the expression developed by Wakelin, (equation 5.1), made possible by advances in data assessment that have recently allowed this data to be re-visited. Introducing the 'y-axis' V_{solid} $G^{0.1667}$ term improves the degree of data fit as shown. This modification derives directly from a detailed consideration of simulated waste solid transport based on later work by McDougall (1995) and has been added

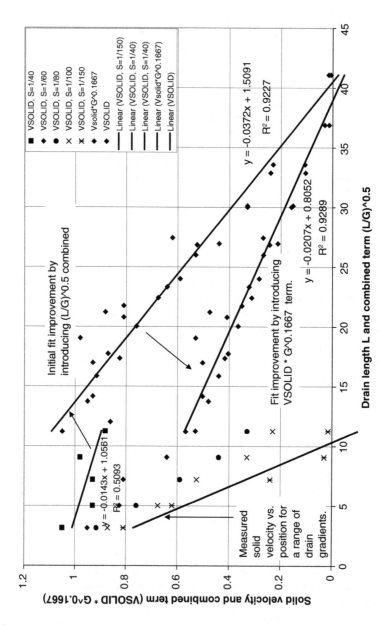

Figure 5.13 Experimental solid velocities for a deformable sanitary product solid discharged to a branch drain set at a range of gradients. Solid velocity dependence on both distance travelled, L, and drain slope, G, is represented by the use of a combined $(L/G)^{0.5}$. A recent minor improvement to the fit based on later simulation studies is also illustrated.

retrospectively to Figure 5.13 as an illustration of both the use of simulations to improve the understanding of solid transport and as a verification of the accuracy of Wakelin's original data analysis.

The recognition that long shallow gradient branch drains, often in inter-floor voids, in new build hospitals in the UK suffered from inordinate deposition problems had initiated Wakelin's research and the follow on site testing undertaken by Bokor from 1980 to 1982 in a large London hospital. Bokor's work (1982) is the only study of faecal waste transport within building drainage systems where wave attenuation is the determinant of solid transport and deposition. This study used an interfloor void drainage network in a large 1960s hospital for a series of site tests over a 2 year period where the installed cast iron branch drains serving 2 w.c.'s each in male and female rest room facilities in a general waiting area, were replaced by UPVC transparent pipe to allow observation of faecal and waste material transport and solid velocity measurement using laboratory proven equipment.

Measurements of solid velocity at 10 locations over a 10 m drain, as well as visual monitoring of flush contents, solid order and dimensions, confirmed that the mechanism of faecal and other waste transport and deposition was identical to that observed in the laboratory for deformable sanitary products. The degree of agreement led to proposals that particular sanitary products could be acceptable as test surrogates for faecal waste. Figures 5.14 and 5.15 present previously unpublished data from Bokor (1984), illustrating multi- and single solid transport and deposition, implying:

- For the single solid cases corroboration of a linear relationship between solid velocity and the $\sqrt{L/G}$ term;
- For the tissue only flushes the zonal description of solid transport is confirmed as in two cases featuring 'thinner' tissue a clear Zone 3 condition is established towards the end of the monitored drain.
- For the multiple faecal solid and tissue case the lead solid velocities are obviously affected as trailing solids close up, increasing the depth differential across the solid and initiating a secondary acceleration, an effect corroborated by laboratory studies, that increased the overall transport distance.

Bokor also identified that the form of the flushed tissue during transport was highly dependent upon the surrounding flow conditions. Initially the tissue was compressed lengthwise by the higher flow velocity close to the head of the drain; however downstream in the more tranquil Zones 2 and 3 the tissue tended to spread out. Bokor observed that tissue transport dominated the final deposition of any leading faecal material – *'tissue transport was almost totally unaffected by interaction with faecal solids while faecal solid transport was greatly influenced by trailing tissue'* – a result later corroborated in laboratory testing (Littlewood and Butler 2003). Bokor identified solid mass as the predominant transport determinant, a conclusion corroborated by dimensional analysis of the relevant system parameters.

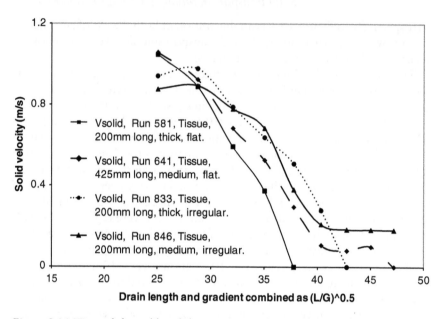

Figure 5.14 Tissue deformable solid transport in a hospital interfloor void branch drain set at 1/200 slope.

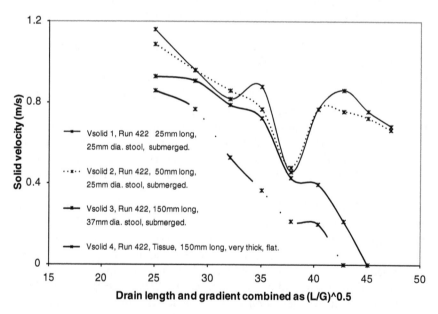

Figure 5.15 Faecal and tissue deformable solid transport in a hospital interfloor void branch drain set at 1/200 slope, illustrating the importance of tissue as a trailing solid in the continuation of faecal solid transport, previously unpublished data from Bokor (1984).

5.3 Simulation of solid transport in attenuating flows

The experimental theme as discussed allows interpolation for cases covered by the particular conditions represented by the test data. For example the results presented here are dependent upon the w.c. type used to insert the test solids. In order to predict solid transport it is necessary to return to the fundamental description of the process involved in transport in order to build a force model that can contribute to the solid–water boundary condition.

Figures 5.16 and 5.17 illustrate a typical solid in transport and identify the forces acting on the solid.

The net force acting on the solid at any point during its motion through the network may be expressed as

$$F_{BODY} = F_{H,up} - F_{Hdown} + (mg - F_B)\sin\theta - f_{fac}(mg\cos\theta - F_B)$$
$$+ \tau_{solid/water} lP_{solid/water} \tag{5.13}$$

where f_{fac} is the solid to wall friction factor, either static applicable prior to motion, or sliding (Swaffield 1982), $P_{solid/water}$ is the wetted perimeter of the

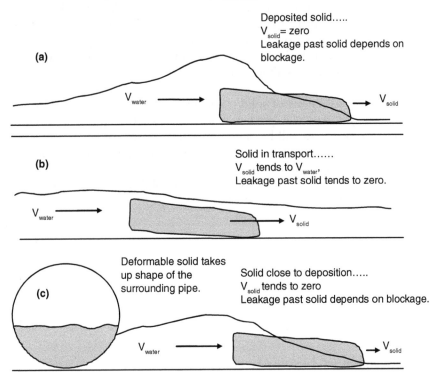

Figure 5.16 Stages of solid transport from inception of motion to subsequent deposition. Note that solids ejected into the flow by a w.c. discharge are assumed to be described by the flow conditions (b) above.

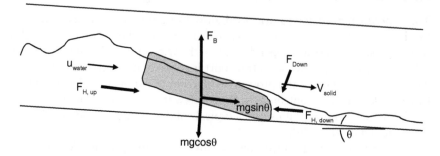

Figure 5.17 Forces acting on a solid during inception of motion, subsequent motion and deposition.

solid and $\tau_{solid/water}$ is the shear stress between the water and the solid. Note that the applicable velocity term in determining the shear force acting on the solid will be $(u_{water} - V_{solid})$ abs $(u_{water} - V_{solid})$ which ensures that when $u_{water} > V_{solid}$ the shear force aids motion but when $u_{water} < V_{solid}$ the shear force acts to retard motion. This effect is clearly visible when a solid is ejected at relatively high velocity from a w.c. due to partial w.c. blockage resulting in a rapid deceleration at the head of the drain and possible deposition prior to re-acceleration by the remaining flush discharge. Values of τ are not readily available.

Based on observations of solid transport it is likely that the hydrostatic and buoyancy forces will predominate, particularly in the final Zone 3 transport. Shear forces will be important in the initial Zone 1 deceleration/re-acceleration zone. Buoyancy will be of the most importance in Zone 2 as the solid reacts to the attenuating flow.

If $F_{BODY} > 0$ then motion is initiated and the solid velocity at the end of a computational time step Δt is given by

$$m\frac{\Delta V_{solid}}{\Delta t} = F_{Body} \tag{5.14}$$

where m is the solid mass and

$$\Delta V_{solid} = V_{solid}^{t+\Delta t} - V_{solid}^{t} \tag{5.15}$$

and the value of solid velocity becomes

$$V_{solid}^{t+\Delta t} = V_{solid}^{t} + \frac{\Delta t}{m}F_{BODY} \tag{5.16}$$

and the distance travelled by the solid may be expressed as

$$x_{solid}^{t+\Delta t} = x_{solid}^t + 0.5(V_{solid}^{t+\Delta t} + V_{solid}^t)\Delta t \tag{5.17}$$

The hydrostatic forces, $F_{H,up}$, $F_{H,down}$ and F_D may be determined from the flow depth upstream and downstream of the solid and the thickness of the solid. (Laboratory and site observations suggested that the F_D downforce could be ignored.) The buoyancy force is determined from an estimate of the submerged solid volume, based on upstream and downstream depths. The gravitational and frictional resistance forces are determined from the solid saturated mass and the static or sliding friction factor assumed; in the cases discussed later values of 0.3 and 0.1 were utilised.

The flow depth upstream of the solid may be determined in terms of the water leakage past the solid at any combination of water and solid velocities. Experimental work indicated that the leakage flow could be defined in terms of the flow specific energy relative to the moving solid – the deposited case is one special case with $V_{solid} = 0$. The leakage flow may be defined as

$$Q_{leakage} = (u_{upstream} - V_{solid})A_{upstream}$$

$$= K_{leakage}\left(h_{upstream} + \frac{(u_{upstream} - V_{solid})^2}{2g} - SE_0\right)^2 \tag{5.18}$$

where SE_0 is the flow specific energy to initiate flow past the solid at rest and $K_{leakage}$ is an empirical coefficient dependent upon solid and pipe cross-sectional areas.

To determine the flow conditions across the solid, either in motion or deposited, there are five unknowns, namely flow depth and velocity upstream and downstream of the solid and the solid velocity at the start of the new time step. The available equations are the continuity of flow equation across the solid, expressed as

$$(u_{p,upstream} - V_{solid})A_{upstream} = (u_{p,downstream} - V_{solid})A_{downstream} \tag{5.19}$$

and the two C^+ and C^- characteristic equations, expressed as

$$u_{p,upstream} = u_R - \frac{g}{c_R}(h_{p,upstream} - h_R) - g(S_R - S_o)\Delta t \tag{5.20}$$

$$u_{p,downstream} = u_s + \frac{g}{c_S}(h_{p,downstream} - h_R) - g(S_S - S_o)\Delta t \tag{5.21}$$

together with the leakage expression 5.18 and the force equation 5.14.

Iterative solution at the solid position yields both flow conditions and solid velocity. Figure 5.18 illustrates the characteristics available and the trajectory of the solid, expressed as a pseudo-characteristic, B⁺, drawn in the x-t plane and representing the track of the solid and having a slope $1/V_{solid}$. In supercritical flows the solid may generate an upstream hydraulic jump whose position and sequent depths are determined as for a normal jump upstream of a junction or flow obstruction (Swaffield 1982). It will be appreciated from Figure 5.18 that the solid position will in all probability always be between nodes so that it is necessary to interpolate to determine the characteristic base values at R and S, a technique already discussed in terms of hydraulic jump modelling.

Solution of the five necessary equations yields a solid velocity and hence a predicted position for the solid at the end of the next time step, equation 5.18. The solid track through the x-t plane describing a single branch or a whole network may then be constructed, as illustrated by Figure 5.19; the figure also illustrates the standard solution for the characteristics at nodes unaffected by the presence of the solid. Solids that remain deposited display B⁺ tracks vertical in the x-t plane as shown. Solid motion is then indicated by a non-vertical track, reverting to the vertical if subsequent re-deposition occurs. Solid transport through junctions may be handled by the technique described as it is probable that the final calculation upstream of the junction will project the solid forward to the downstream branch.

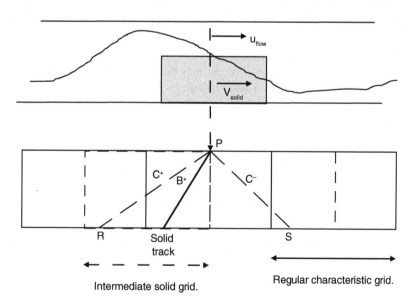

Figure 5.18 Development of the solid transport characteristic solution and identification of the solid track.

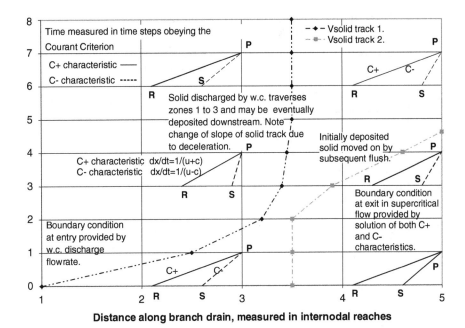

Figure 5.19 Method of characteristics solution for unsteady free surface flows, including the representation of solid track within the x-t plane.

In the case of deposited solids the flow downstream of the solid may be taken as at Critical depth with a flow velocity equal to the local wave speed. Similarly if the simulation is of a solid discharged to the branch drain from a w.c., then the initial solid velocity may be set equal to the w.c. discharge profile velocity at the head of the branch. The solution technique will be identical. Early discharge from a w.c. may result in almost immediate deposition as the set velocity is too high to be supported over the next time steps by the flow depths encountered. This is a result consistent with the Zone 1 action identified by Wakelin (1978).

Figure 5.20 presents a comparison between the simulation described and the observed transport of a single solid from rest along a 14 m length of 100 mm diameter UPVC branch drain subject to a w.c. discharge with the profile illustrated. The model solid was an 80 mm long, 37 mm diameter cylinder with a specific gravity of 1.10. Buoyancy forces were calculated on the assumption of a depth profile along the solid from upstream to downstream face – linear and elliptical profiles were considered. As described, solid velocity was recorded by means of the standard photo-electric cell instrumentation (Swaffield and Galowin 1992). Depth measurements were made as the solid traversed the branch drain and are included in Figure 5.20. Figure 5.21 illustrates the form of the depth profiles across the solid routinely

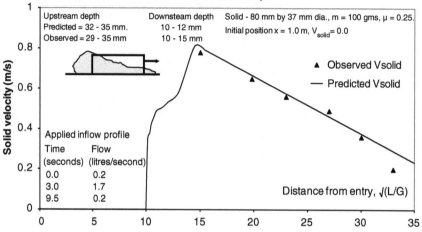

Observed and predicted solid velocities along a 14 m, 100 mm diameter UPVC drain set at a slope of 1/80.

Figure 5.20 Comparison of observed and predicted solid transport from rest using the boundary equations developed to describe the forces acting on the solid and the leakage flow past the solid.

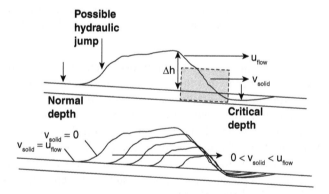

Figure 5.21 Flow depth profiles observed along the length of a cylindrical solid as the solid velocity increases from zero to the local flow velocity.

observed during such tests – note that once the solid velocity reaches parity with the local flow velocity the depth differential tends to zero.

The boundary equations detailed in equations 5.14 to 5.21 allow the velocity of the solid relative to the surrounding water to be simulated, together with the depth differential across the solid during transport. Figure 5.22 illustrates such a simulation for the transport of a 250 gm saturated mass sanitary pad and also provides a comparison to a 125 gm saturated pad identical in all

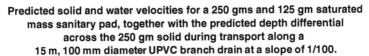

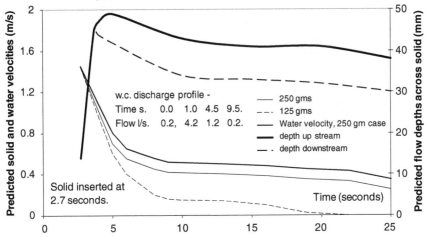

Figure 5.22 Comparative solid transport velocity predictions for two pads differing only in base area and saturated mass. The depth differential across the larger pad is also illustrated to confirm the applicability of the force boundary conditions.

respects except for the pad width which is reduced from 70 mm to 35 mm. It will be seen that the transport of the smaller pad is less efficient; a result that confirms the dimensional analysis represented by equations 5.10 and 5.11. It will be seen that the solid width, w, appears in groups π_3 and π_7; however in the latter case the reduction in w has no effect as the mass is also halved. Thus the effect will be seen in the influence of π_3, as shown on Figure 5.12. Taking a datum value of π_3 based on a 5% blockage area in a 100 mm diameter drain and comparing that to the per cent blockage represented by solid base areas of 70 by 20 mm² and 35 by 20 mm² indicates that the π_3 axis value moves from 1.4 to 2.8 for the narrower and wider pads respectively, with a consequent reduction in C_3 relative to a 5% blockage datum from 0.7 to 0.4. Thus the dimensional analysis reinforces the simulation conclusion that a marked reduction in hydrostatic forces due to the reduced base area reduces the likely transport distance for this solid.

The successful modelling of discrete solid transport was taken further by a consideration of the transport of floating solids where it was possible to accurately model the transport by application of a velocity decrement factor applicable between the solid and the surrounding water. Figure 5.23 illustrates an early application where a constant decrement factor, in the three cases illustrated set at 1, 0.9 and 0.8 sequentially, was seen to yield

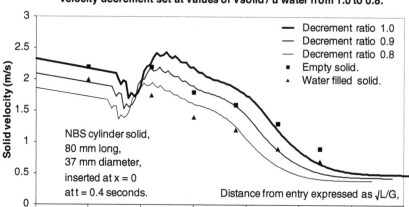

Figure 5.23 Solid velocity simulation derived from floating solid observations using a constant velocity decrement factor.

solid velocities at least of the same order as those recorded during laboratory measurements. Velocity decrement laboratory observations for the transport of latex water filled sheath solids, 25 mm diameter, 100 mm long, developed at Stevens Institute, Hoboken, as possible w.c. discharge test solids, in a steady flow (Swaffield and Galowin 1992), are presented in Table 5.3.

5.4 Interacting solids

A significant feature of solid transport is how multiple solids interact with each other. The presence of other solids causes changes in the water depth profile of the flow in the pipe, such that they speed up and slow down as they interact with each other. This spring-like motion is considered important, particularly in the deposition region as solids can travel considerably further as a result.

The transport of a solid in a near horizontal drainage pipe under steady flow conditions is characterised by a number of significant changes in the flow depth profile along the pipe, as shown in Figure 5.24. The water height behind the solid reduces gradually to a point where the water depth is normal for the particular flow regime due to the inflow and the water immediately in front of the solid is below Normal water depth and increases downstream. This bow wave is due to the effects of water tumbling over the solid at a higher than normal velocity, as shown in Figure 5.24.

Table 5.3 Velocity decrement laboratory observations for the transport of latex water filled sheath solids

Solid description	Solid sg.	Drain slope	Applied flowrate, (litres/second)	Average steady flow velocity (m/s)
NBS 80 mm by	0.85	1/100	2.0	0.83
37 mm diameter solid	1.05		1.5	0.79
	1.23		1.0	0.72
NBS 80 mm by	0.85	1/50	2.0	0.96
37 mm diameter solid	1.00		1.0	0.86
	1.05		0.67	0.76
	1.23		0.5	0.67
NBS 80 mm by	0.85	1/100	1.5	0.79
25 mm diameter solid	1.00		1.0	0.72
	1.05		0.83	0.68
	1.23		0.67	0.63
NBS 37 mm by	0.85	1/50	2.0	0.96
37 mm diameter solid	1.05		1.0	0.86
	1.23		0.67	0.76
NBS 37 mm by	0.85	1/100	2.0	0.83
37 mm diameter solid	1.00		1.5	0.79
	1.05		1.0	0.72
	1.23		0.55	0.57
NBS 80 mm by	0.85	1/50	2.0	0.96
25 mm diameter solid	1.00		1.0	0.86
	1.05		0.50	0.67
	1.23		0.33	0.75
Stevens Institute latex	0.83	1/100	1.25	
solid, painted	1.04		1.0	0.72
	1.10		0.67	0.63
Stevens Institute latex	0.83	1/100	1.25	
solid, unpainted	1.04		1.0	0.72
	1.10		0.67	0.63
Stevens Institute latex	0.83	1/50	1.5	0.92
solid, painted	1.04		1.0	0.86
	1.10		0.5	0.67
Stevens Institute latex	0.83	1/50	1.5	0.92
solid, dye filled	1.04		1.0	0.86
	1.10		0.5	0.67

5.4.1 Solid velocity measurement

Velocity measurement of a solid is not straightforward. At the outset of this investigation the available technology was found lacking. Velocity measurement techniques used in industrial applications rely on the uniform movement of the object to be measured, and small distances between the object and the pickup device. For solids travelling in partially filled pipes under unsteady flow conditions the situation could not be more different.

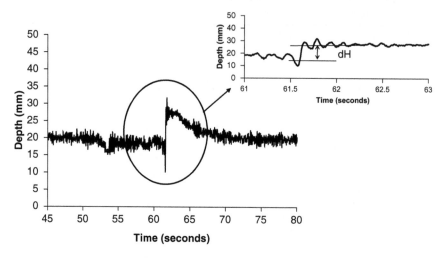

Figure 5.24 Water depth difference across a solid.

Systems used in the past by the Drainage Research Group at Heriot-Watt University and others earlier (McDougall 1995) relied on the solid breaking a visible light beam. When the beam is broken a signal is registered and a time logged. There are several problems with this approach:

- The system uses visible light so it is prone to false triggering from surrounding light sources (sunlight, artificial light).
- The movement of a solid through the water creates waves which cause reflections and can 'confuse' the system (Wakelin 1978).
- Because solids in motion are subject to movement anomalies (see below) it is possible for the system to completely miss a solid.
- The system is not capable of detecting solids at the extremes of the flow rate range – i.e. flow rates less than 0.8 l/s and flow rates above 4 l/s (in some cases less).

Previous data gathered on solid velocities (McDougall 1995; Clark 1994) have been shown to be accurate; however the limitation of the range of flow rates, particularly at the lower flows, and the time taken to achieve the results suggested that an improvement was required. The low flow rate data is of particular interest in this research as it is in this region that interaction is likely to occur.

5.4.2 Solid movement anomalies

Assuming the solid in question to be a rectangular cylinder in plan, the object is subject to three main movement anomalies as it travels in a flow:

- Pitch: The solid pitches from front to back.
- Roll: The solid rolls.
- Yaw: The solid swings from side to side.

The difficulty arises in how to ensure that the method of pickup used can cope with these differing distances between object and pickup and angle of approach.

5.4.3 The multiple solid position detection system

In response to the problems encountered with repeatability of solid velocity measurements in the past, the Multiple Solid Position Detection System (MSPDS) was developed. The system uses an active solid fitted with an infrared emitter. As a solid passes a pre-defined location along a pipe length an infrared detector is activated and produces a digital pulse. This pulse advances a counter giving an indication that detection has occurred; it also shows how many solids have passed that location in the case of multiple solids. The detection card then channels the output pulses from the detector so that all outputs for the first solid are input to the computer via channel 1 and the data for the second solid are input into channel 2 and so on. The output of data onto different channels effectively identifies each solid and makes future processing in a spread sheet easier. The system has the added advantage of requiring only the same number of channels as solids (typically 1 to 4), rather than the number of detectors (typically 28 to 40). The arrangement is shown in Figure 5.25.

The enhanced resolution of the system provides the opportunity to look at solid transport in general and the interaction of solids in particular.

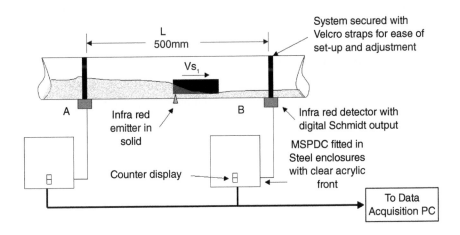

Figure 5.25 Solid velocity measurement.

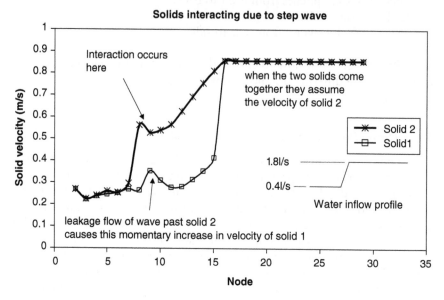

Figure 5.26 Interacting solids.

Figure 5.26 shows a trace of the interaction of solids in a 100 mm pipe. Solid 1 is allowed to enter the pipe on a small baseflow of 0.4 l/s, enough to sustain movement. When solid 1 has passed 2 m, solid 2 is entered. When solid 2 reaches 1 m a surge is initiated up to 1.8 l/s. Solid 2 accelerates until it enters the backwater of solid 1 where it decelerates until both solids are swept away by the surge. It can be seen from Figure 5.26 that the developed measuring system is capable of recording this interaction, which occurs over a relatively short time and distance.

5.4.4 Solid boundary condition

The presence of the solid in a flow requires a simple modification to the water depth profile at the solid's location along the pipe at a given time. The water depth is therefore given by

$$h_i = h_i + dH_S \tag{5.22}$$

where h_i is the water depth at node i if no solid is present and the flow can be either supercritical or subcritical, and dH_s is the water depth difference across the solid which is a function of the velocity of the solid, V_s such that;

$$dH_S = f(V_S) \tag{5.23}$$

A gradually varied profile can then be fitted between the solid boundary and the upstream location where the flow depth returns to h_r. It should be noted that the boundary condition given by equation 5.23 is valid for a solid at one location only or for the special case where a solid has deposited.

5.4.5 A mathematical model based on dH_s

The classification of the movement of a solid in a flow based on the water depth difference across the solid presents a simple means of defining solid velocity in steady supercritical flows. The model however needs to be able to cope with solid transport in a variety of flow regimes. The following describes a general mathematical model for the transport of solids in all flow regimes where a water depth difference exists. While the classification is for water depth difference across the solid, it was found that a more robust model was obtained when the water depth was classified in terms of the wave speed of the flow at that depth. This leads to the general classification of solid transport in terms of the Mach Number of the solid thus:

$$M_s = \frac{V_s}{c_n}$$

(5.24)

The water depth difference dH_s is related to V_s in that dH_s is at its maximum when $V_s = 0$, and dH_s is at its minimum when $Vs \to V_{s(max)} \to V_f$. In keeping with the general classification of water depth in terms of wave speed then the following can also be said of the solid:
when $Vs \to 0$

$$\frac{c_{us}}{c_{ds}} \to \frac{c_{us}}{c_{ds}} max$$

and when $Vs \to V_{s(max)} \to V_f$

$$\frac{c_{us}}{c_{ds}} \to 1$$

where c_{us} is the wave speed at the upstream face of the solid
and c_{ds} is the wave speed at the downstream face of the solid.
The 'positive' curve shown in Figure 5.27 shows the relationship between c_{us}/c_{ds} for a single solid of diameter 36 mm, specific gravity 1.05 in a 100 mm pipe set to a slope of 1 in 100. The trendline shown represents the possible acceleration or deceleration of a solid when $c_{us}/c_{ds} \geq 1$, that is when there is a water depth difference across the solid and the solid is not fully buoyant. The data for this curve was obtained under normal supercritical conditions;

therefore $c_{ds} \approx c_n$. In practice there is likely to be a depth depression down-stream of the solid, so it is usual for the following expression to be true:

$$c_{ds} \leq c_n \qquad (5.25)$$

where c_n is the wave speed of the flow at Normal depth.

So the positive curve in Figure 5.27 is applicable when the condition in equation 5.25 is met. In the case where a solid is travelling in a flow where the flow is not supercritical, then the inverse curve can be used. In this case the following condition applies:

$$c_{ds} \leq c_n \qquad (5.26)$$

This equation will cause a solid to decelerate as it enters deeper slow mov-ing water such as the backwater of another solid.

The measure of success for this method of modelling solid transport in drain pipes can be described as follows;

1 A water depth difference across a solid can be made to move along a pipe at a velocity dependent upon the characteristics of the solid.
2 The velocity of a solid is modified as it enters the backwater of another solid.
3 The water depth difference of a solid varies as its velocity varies.
4 A single solid will deposit when the water conditions dictate in the pipe; however the stationary solid will allow a water build-up behind it insti-gating movement again; this process can repeat, nudging the solid fur-ther than predicted by velocity decrement models.

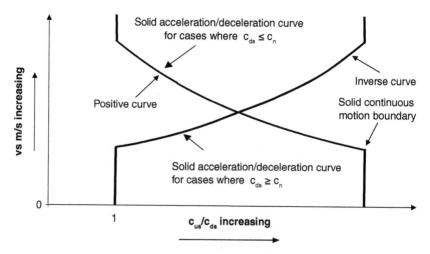

Figure 5.27 General form of the model.

5   The distance between solids varies as they travel along a pipe subject to an attenuating wave in a push–pull fashion until deposition finally occurs.

Figure 5.28 shows a water depth history along a pipe when a solid is present. The flow is steady and the water depth associated with the presence of the solid is seen to move across the nodes.

Figure 5.29 shows the modification of solid velocity as a solid moves into the backwater of another solid. Figure 5.29 also shows the model's ability to modify the water depth difference across the solid as the solid velocity varies. At the start of the test run when the solids are travelling in a steady flow both solids have a steady dH_s. The acceleration of solid 2 due to the arrival of the surge causes dH_s for solid 2 to decrease. Solid 2 decelerates as it moves into the backwater of solid 1 and as a consequence, dH_s increases again until the two solids have moved close together. At this point both solids follow a similar solid velocity down the pipe and the value of dH_s settles down to a common level. It is interesting to note though that even in this relatively steady flow there is still a push–pull element to the travel, particularly for solid 2 as it travels in and out of the backwater of solid 1.

Figure 5.30 shows a predicted velocity profile for a solid subject to the inflow profile that is also shown. The solid velocity decreases with the attenuating wave as it travels along the pipe. When the velocity of the solid falls below 0.2 m/s, there is a levelling off and an eventual deposition. This levelling off represents transport in zone 3 as defined by Wakelin (1978) and is the velocity measured between nodes. The solid is seen to deposit and start moving again due to the water build up behind it. This process repeats until a final deposition occurs.

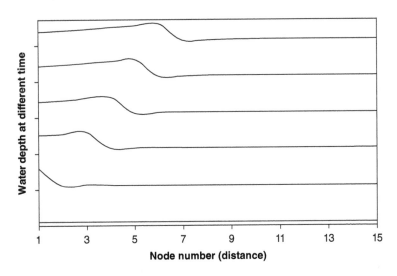

Figure 5.28 Water depth history along pipe showing dH$_s$.

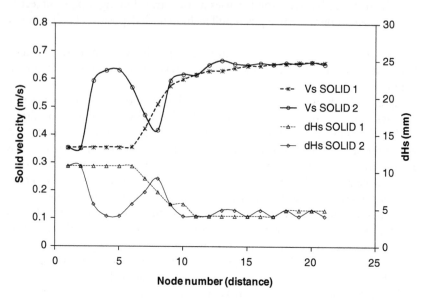

Figure 5.29 Variation in dH$_S$ with V$_S$ when solids interact.

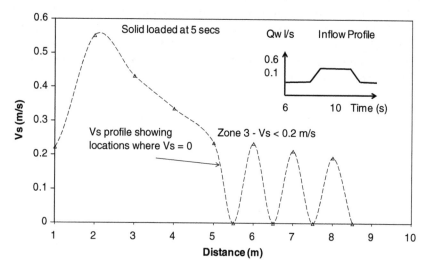

Figure 5.30 Solid deposition showing stop/start motion.

The case for the deposition of multiple solids is similar; however the effects of their interactions are also significant. Figure 5.31 shows the variation in the distance between solids as they move along a pipe. Solid 1 is input at 1 second, and solid 2 is input at 5 seconds. The solids are subjected to a surge wave as shown on the inflow profile on Figure 5.30. Solid 2 is affected by the

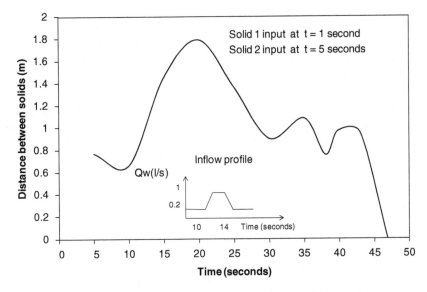

Figure 5.31 Variation in distance between solids as they travel along pipe.

wave first, moving it closer to solid 1. The effects of the proximity to solid 2 and the arrival of the surge wave pushes solid 1 forward, moving the solids further apart. As the wave abates both solids approach each other again, interacting in a push–pull fashion until the wave and the effects of interaction abate and the solids deposit.

This mach model with interaction has been shown to accurately predict the motion of multiple solids in near horizontal building drainage systems. Existing models deal well with the calculation of solid velocities based on the surrounding water conditions but do not modify them; this seriously limits the flexibility of the system in that equation are required for every scenario the solid encounters. This method provides a more robust deterministic approach, where depth profiles are modified from which other variables can be calculated. The model has been shown to effectively predict the interaction between the velocity of the solid and the upstream water depth in a variety of scenarios, including interaction between solids.

5.5 Concluding remarks

This chapter has set out the extensive work carried out in the area of solid transport in building drainage systems. The ability to include solid movement in the Method of Characteristics model DRAINET provides a powerful tool with which to analyse system performance. A sound understanding of the mathematical nature of solid movement in a flow provides a basis for much of the theoretical work carried out over the years. This chapter has placed solid transport in its historical setting and has evaluated solid transport in horizontal pipes using steady flows, testing regime using w.c. discharges and laboratory investigation using innovative sensing technology.

The validation of the mathematical modelling proposed here is essential since the credibility of any possible policy making centres on the ability to demonstrate that the models are appropriate and robust.

Innovative solutions, such as the use of non-circular pipes, which turn out to originate from the nineteenth century are evaluated as are the influences of 'deformable solids' which describe much of the waste found in building drains and interacting solids; all contribute to the overall confidence in the methods employed and the techniques used to provide the extent of system analysis possible.

Note

1 PERC has since reported its findings on drainline carry and the influence of w.c. flush parameters. In general the report concludes that flush volume is an important parameter and concludes that large 'cleansing' volumes have limited application due to wave attenuation. The team's approach was thorough but led to the conclusion that trailing volume is not a significant factor in solid transport. The decision to make a very long test rig (40 m) forced an averaging that led to this conclusion. The work has played a significant role in raising the profile of 'drainline carry' and is to be commended in this regard.

6 Rainwater drainage systems

This chapter details the application of flow modelling to rainwater drainage systems, both conventional and siphonic. It starts with a brief overview of the operation of rainwater drainage systems, before going on to outline possible approaches to the simulation of conditions within such systems. Throughout, the underlying principles are illustrated through application of some of the many numerical models that have been developed over the past 20 years at Heriot-Watt University (FM4GUTT, SIPHONET, ROOFNET).

6.1 Introduction

As shown in Figure 6.1, there are basically two types of rainwater drainage system. The pipework within conventional systems operate at atmospheric pressure, and the driving head is thus limited to the gutter flow depth, whereas that within siphonic systems is designed to run full bore, resulting in sub-atmospheric system pressures, higher driving heads and higher flow velocities. As a result, siphonic systems normally require far fewer downpipes to drain the same size roof area, and the depressurised conditions also mean that much of the collection pipework can be routed at high level, thus reducing the extent of costly underground pipework. These advantages make siphonic systems ideally suited to large industrial and commercial roof areas, with high profile examples including the Sydney Olympic stadium and Hong Kong Airport, as well as historically Stansted Airport, Edinburgh Murrayfield Stadium and many other commercial applications.

To generate the full bore conditions necessary for siphonic action, a siphonic system utilises specially designed gutter outlets that eliminate air entrainment once the gutter water levels have achieved a certain level (Figures 6.2 and 6.3). Any storm below the design threshold will result in oscillatory free surface/unsteady full bore flow conditions within the system, whilst storms which exceed the design condition will result in flooding unless excess run-off is re-directed. Accordingly, the main disadvantage associated with siphonic systems is that, for any network, priming requires a particular storm intensity. Another disadvantage is the need to ensure that siphonic conditions are not lost in multiple outlet systems if one or more outlets admits air

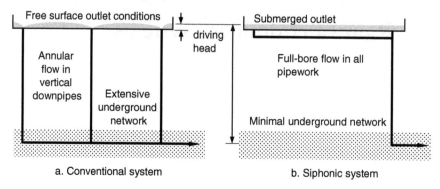

Figure 6.1 Schematic of conventional and siphonic rainwater drainage systems.

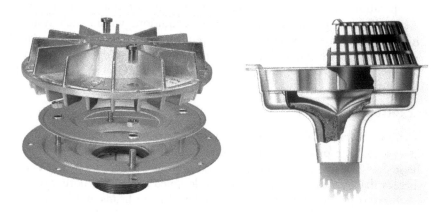

Figure 6.2 Typical siphonic roof outlets. (Source: (left) Copyright (c) 2000 UV-System. (right) Copyright (c) 2000 Fullflow Limited, by kind permission of Fullflow Group Limited.)

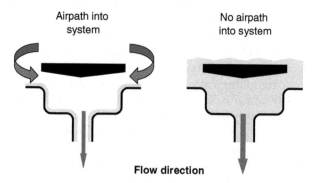

Figure 6.3 Typical gutter outlet illustrating entrained airflow exclusion baffle.

and hence breaks the siphon. This implies that multiple outlet networks must be balanced so that all gutter outlets remain submerged. Below 40% of the primed system capacity, single outlet systems act as conventional rainwater drainage, while above this level oscillatory unsteady flow conditions prevail.

Both conventional and siphonic systems consist of three interacting components (roof surface, collection gutters, system pipework), and each of these components has the ability to substantially alter the runoff hydrograph as it passes through the system. As shown in Table 6.1, the prevailing flow conditions through each of these basic components are fundamentally different.

6.1.1 Roof surfaces

Notionally, a roof may be described as flat, sloped or green. Flat roofs are defined as those below a certain gradient, e.g. 10% in the UK (BSI 2000), and they are commonly used with domestic properties in climates with low rainfall or with industrial buildings in developed countries. Most residential and many commercial properties in the developed world have sloped roofs, as their ability to drain naturally means that there is less risk of leakage.

In addition to traditional roof surfaces (e.g. tiles, sheets, EDPM), green roofs have also started to become more prevalent in recent years. A green roof is essentially a roof that incorporates an upper layer of vegetation and some form of growing medium, along with drainage, filtration and waterproofing layers (see Figure 6.4). Although they are installed for a variety of different purposes, they are most often associated with Sustainable Urban Drainage System (SUDS) to help manage stormwater. One of the underlying principles behind SUDS is the concept of source control, whereby rainfall is dealt with where it lands rather than passing it downstream where it can contribute to increased flood risk. The use of green roofs represents one such technique; rainfall can either be stored within plants and substrate before evapotranspiration, or attenuated to some degree as it passes through the various layers that make up the roof structure. These two processes can result in significant reductions in both peak runoff rates and total runoff

Table 6.1 Prevailing flow conditions in a rainwater drainage system

System component	Flow type	Inflows	Flow dimension	Flow depth relative to width
Roof	Free surface	Laterally varying	2D	Very low
Collection gutters	Free surface	Laterally varying	Can be approximated as 1D	Low to comparable
System pipework	Annular, full bore, pulsing, two phase	Point	Can be approximated as 1D	Low to comparable

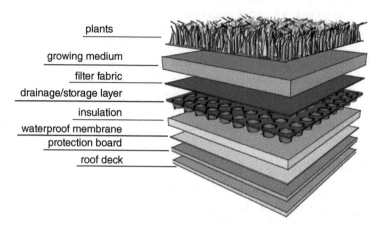

plants
growing medium
filter fabric
drainage/storage layer
insulation
waterproof membrane
protection board
roof deck

Figure 6.4 Typical composition of a green roof.

volumes. Green roofs can also offer other benefits, or ecosystems services (ESS), including: increased urban biodiversity, decreased urban heat island effect, improved air quality, improved runoff water quality, reduced building energy use, improved acoustic insulation, improved amenity and increased roof lifespan. The ESS provided by green roofs depends primarily on the depth of substrate and the vegetation used. *Extensive* green roofs have shallow substrates (50–100 mm) and, whilst they can only support drought-tolerant plants such as sedum, their low weight and cost means they are ideal for new or retro-fit applications. *Semi-intensive* and *Intensive* green roofs have deeper substrates, can support a wider range of plant types and, whilst their higher cost and weight renders them less widely applicable, they can provide enhanced ESS over extensive roofs. *Biodiverse* roofs are similar in composition to extensive roofs but differ in growing medium and vegetation, which are selected to replicate/enhance the original habitat of an area.

6.1.2 Collection gutters

The basic requirement for collection gutters is that they have sufficient flow capacity to accommodate flows from the design storm. Flow conditions within gutters are typically subcritical, due to their negligible longitudinal gradient, and incoming flows tend to split evenly between adjacent gutter outlets (see Figure 6.5). Water surface profiles will slope towards the outlet, and it is the difference in hydrostatic pressure along the gutter that gives the incoming water the required momentum to flow towards the outlet (May and Escarameia 1996). Key to ensuring whether or not collection gutters have sufficient capacity are the conditions that occur at the gutter outlets. As well as affecting the flow rates entering the drainage system pipework, the outlet depths also affect the upstream gutter depths (via the water surface profile). Hence, although the depth at a gutter outlet may not cause any

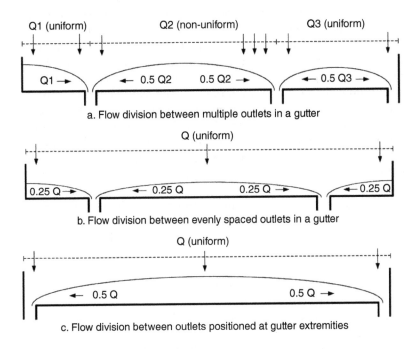

Figure 6.5 Typical gutter water surface profiles.

particular problems, the higher depths occurring at the upstream end of the gutter may result in overtopping. Extensive experimental studies in the 1980s determined that the flow conditions in the vicinity of a gutter outlet in a conventional rainwater drainage system could be categorised as being either 'weir' type or 'orifice' type, depending on the depth of water relative to the size of the outlet (May 1995). In siphonic systems, the outlets are designed to become submerged in order to allow full bore flow conditions to develop and be sustained. If this is the case, the outlet depth is dependent upon the downstream conditions within the connected pipework as well as the gutter inflows. Previous experimental work has indicated that siphonic action may also occur for a short distance below the outlet in conventional rainwater drainage systems incorporating 'wide gutters' (Wright *et al.* 2006), which allow more water to reach the outlet. In this context, wide gutters are defined as those where the gutter sole width is at least twice the diameter of the gutter outlets, whereas 'standard' gutters are defined as those where the gutter sole width is ~ equal to the outlet diameter.

6.1.3 System pipework

The type and extent of pipework incorporated into a rainwater drainage system depends primarily on whether the system is conventional or siphonic.

In conventional systems (Figure 6.1a), the above-ground pipework generally consists of vertical downpipes, connecting the gutter outlets to some form of underground drainage network. System conditions are normally free surface and, if the downpipes are sufficiently long, annular flow conditions may occur. System capacity is usually dependent upon the capacity of the gutters rather than the capacity of the vertical downpipes. However, the use of wide gutters (outlined above) can result in full bore flow within downpipes.

In contrast to conventional systems, siphonic installations depend upon the purging of air from the system (priming) and the subsequent establishment of full bore flow conditions within the pipework connecting the gutter outlets to the surface water network at ground level (Figure 6.1b). Consequently, the flow conditions within a siphonic system may be free surface, full bore or some combination of the two regimes. In order to simulate unsteady flow regimes in a system under all operational conditions it is necessary to understand the cyclic nature of the problem. Figures 6.6 and 6.7 illustrates a two-outlet siphonic system and the mechanisms required for priming thus (Wright *et al.* 2006):

1   Free surface flows in both the gutter and the system pipework with annular flow in the vertical stack.
2   As flow rate increases a hydraulic jump forms upstream of the pipe junction in both pipes.
3   Increasing inflow leads to full bore flow at the junction that propagates downstream towards the vertical stack.
4   When the full bore flow reaches the stack, de-pressurisation of the system occurs, increasing the inflow to the system from the gutter outlets

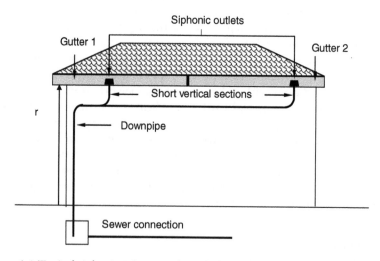

Figure 6.6 Typical siphonic rainwater drainage system.

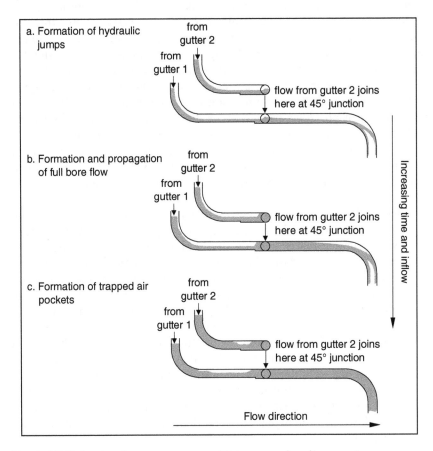

Figure 6.7 Siphonic rainwater system establishment and cyclic operation.

and the establishment of full bore flow at the head of each branch – trapping air pockets in both pipes.

5   Both air pockets are swept downstream. Air pockets that reach the vertical stack cause a momentary re-pressurisation until the air exits the system at the stackbase.

6   Once the air has exited the system pressure remains steady, and the system operates at its design condition.

During the priming phase of a system, or periods when the rainfall intensity falls below the design point, the conditions within a siphonic system may well contain substantial quantities of entrained air and exhibit pulsing or cyclical phases, a result of greatly varying gutter water levels and an indication of truly unsteady, transient flow conditions. Such conditions can be problematic and are exacerbated when the system incorporates more than one outlet connected to a single downpipe, as the breaking of full bore conditions at one of the outlets (due to low gutter depths and air entry) is transmitted throughout the system and, irrespective of the gutter depths above the remaining outlet(s), results in cessation of fully siphonic conditions. It is normally during sub-design events that the majority of minor operational problems tend to occur, e.g. noise and vibration. More serious problems include inadequate system capacity and the actual implosion of pipework due to very low system pressures; such problems usually occur due to a combination of relatively high rainfall and human error (design, installation or maintenance defects).

Previous research has shown that, depending on system inflows, a siphonic system incorporating two gutter outlets will display the following flow regimes as rainfall intensity rises to the design level (Wright *et al.* 2006):

- Regime 1 – system inflows up to 40% of the design criteria inflows: highly unsteady conditions, characterised by cyclical periods of positive and negative pressures caused by low flow depths in one or both of the gutters so that siphonic action could only be sustained for short periods. Siphonic action would quickly drain the gutter(s), creating an airpath to the atmosphere and hence breaking the siphon and reducing system flow rates, allowing gutter flow depth to increase until siphonic action was re-established.
- Regime 2 – system inflows between 40% and 60% of the design criteria inflows: oscillating, constantly negative system pressures above those associated with the fully primed system caused by intermediate flow depths in one or both gutters being sufficiently high to ensure a continuous siphonic action but not deep enough to 'swamp' the vortices that occur around the outlets and entrain air into the water flow, leading to lower flow rates and higher pressures than those associated with the fully primed system.
- Regime 3 – system inflows above 60% of the design criteria inflows: system pressures initially mirror those in a fully primed system, although tending to return to the oscillatory pressures associated with Regime 2. Such conditions arise if the flow depth in one or both of the gutters was only sufficient to sustain full siphonic action for a short period after which the gutter depth decreased to allow large quantities of air to become entrained with the water inflows.

6.1.4 Current design methods

The design of the roof surface is usually within the remit of the architect rather than the drainage designer, highlighting that structural and/or aesthetic concerns often take precedence over performance criteria. The hydraulic design of gutters is complicated by the presence of the sloping water surface profile, the precise form of which can only realistically be determined with recourse to some form of numerical model. There are no commercially available models to undertake this level of analysis and, consequently, current design methods for gutters installed in conventional drainage systems are based primarily on empirical relationships (May 1994) and the assumption of free discharge at the outlet. Little additional guidance is given in the standards for the design of gutters in siphonic systems, and the onus is firmly on the system designers to ensure adequate capacity; this is normally ensured by the laboratory testing of specific gutter outlets. With respect to system pipework, the relevant standard (BSI 2000) specifies that downpipes in conventional systems should run no more than 33% full. Similarly, the flow conditions within offset pipes must also normally be free surface, with BS EN 12056-3:2000 specifying that offsets run no more than 70% full. Current design practice for siphonic systems assumes that, for a specified design storm, a siphonic system fills and primes rapidly with 100% water (Arthur and Swaffield 2001). This assumption allows siphonic systems to be designed utilising steady-state hydraulic theory, normally in the form of simple computer programs. The steady flow energy equation is employed (May 1995), with the elevation difference between the gutter water level and the point of discharge being equated to the head losses in the system. However, steady-state design methods are not truly applicable when a siphonic system is exposed to a rainfall event below the design criteria, or an event with varying rainfall intensity. Furthermore, current design methods cannot account for commonly occurring operational problems, such as the blockage of outlets and the submergence of the system exit due to downstream sewer flooding.

6.2 Roof surface

The flow conditions on roof surfaces are normally independent of the type of drainage system, and hence a similar simulation technique can be applied to roofs connected to both conventional and siphonic drainage systems.

6.2.1 Numerical approach

A kinematic wave approach can be used to route rainfall over roof surfaces and into the gutters. Although less accurate than a full dynamic approach, the relatively large surface areas involved mean that anything more complex would result in unfeasibly long computational run times. Combining the continuity and momentum equations for a kinematic wave (Chow *et al.* 1988) yields:

$$\frac{\partial Q_r}{\partial x_r} + \alpha \beta Q_r^{(\beta-1)} \frac{\partial Q_r}{\partial t} = q_r \qquad (6.1)$$

where Q_r = flow rate down roof, x_r = distance down roof (from apex),

$\alpha = \dfrac{n_r p_r^{\frac{2}{3}}}{1.49\sqrt{S_{or}}}$, n_r = roof Manning's roughness coefficient, p_r = roof width,

S_{or} = roof slope, β = roof geometry dependent parameter (0.6 for a rectangular roof), t = time and q_r = flow rate onto roof (dependent on rainfall intensity).

This equation can be transformed into a finite difference representation using a backward-difference linear scheme (Chow *et al.* 1988), as shown below, which is then readily solved for discrete time steps (Δt) and space steps (Δx):

$$Q_{i+1}^{j+1} = \frac{\left[\frac{\Delta t}{\Delta x}Q_i^{j+1} + \alpha\beta Q_{i+1}^j\left(\frac{Q_{i+1}^j + Q_i^{j+1}}{2}\right)^{\beta-1} + \Delta t\left(\frac{q_{i+1}^{j+1} + q_{i+1}^j}{2}\right)\right]}{\left[\frac{\Delta t}{\Delta x} + \alpha\beta\left(\frac{Q_{i+1}^j + Q_i^{j+1}}{2}\right)^{\beta-1}\right]} \qquad (6.2)$$

In addition to variable rainfall conditions and roof geometries (area, slope, roughness), it is also possible to account for the effect of wind-driven rain and vertical surfaces by modifying the roof area subject to rainfall; for example, rain falling vertically at an intensity of 20 mm/hr will result in run-off of 9.8 l/s on a 50 m × 50 m roof at an angle of 45° (roof area exposed to rainfall = 50 × cos 45° × 50 = 1768 m²), whilst rain falling perpendicular to the same roof will result in runoff of 13.9 l/s (roof area exposed to rainfall = 50 × 50 = 2500 m²).

The basic effects of green roof surfaces can also be simulated utilising the Horton infiltration formulation, which calculates the quantity of rainfall that infiltrates into a green roof rather than running off into a gutter thus:

$$f_t = f_c + (f_0 - f_c)e^{-kt} \qquad (6.3)$$

where f_t is the infiltration rate at time t, f_0 is the initial (maximum) infiltration rate, f_c is the minimum infiltration rate and k is a constant based on soil type.

It should be noted that the Horton equation presented implicitly assumes uniform soil infiltration, and therefore cannot be used to account for evapotranspiration or soil moisture variations down a sloping roof surface; in reality such variations are unlikely to be significant during a typical rainfall/runoff event. However, more accurate approaches incorporating these effects are available (Shaw 2008).

Whilst the above approach can be applied to roof surfaces supplying gutters, flat roofs connected directly to downpipes must be treated slightly differently. A simple volumetric approach is probably the most suitable technique, whereby the quantity of rainwater falling on a roof during any given interval is used to calculate the corresponding depth of water on the roof, which then controls the flow rate into the connected downpipe(s), thus:

$$y_{r_i} = y_{r_{i-1}} + I\Delta tA - \sum Q_o \Delta t \qquad (6.4)$$

where y_{r_i} = roof depth at current time, $y_{r_{i-1}}$ = roof depth at previous time, I = rainfall intensity, Δt = timestep, A = roof area, Q_o = outflow from roof (either into collection gutter or through outlets at previous time).

6.2.2 Boundary conditions

The boundary conditions describing flow at the roof outlets depend on the prevailing flow regime. If the flow in the connected downpipe remains free surface, the outlet flow is calculated assuming weir flow (BSI 2000), thus:

$$Q_o = y_{r_i}^{1.5} \frac{D_o}{7500} \qquad (6.5)$$

where D_o = outlet diameter.
  If the flow in the connected downpipe becomes full bore, the conditions on the roof surface are inextricably linked to those within the downpipe, and the available C^- characteristic equation at the entry to the downpipe can be solved with an expression linking roof depth to pipe pressure, thus:

$$P_1 = \rho g y_{r_i} + P_{atm} + \Delta P_o \qquad (6.6)$$

where P_1 = pressure at entry to downpipe, ρ = density of flow (dependent on air content), g = acceleration due to gravity, P_{atm} = atmospheric pressure, ΔP_o = pressure loss through roof outlet.
  If the roof outlet becomes totally blocked by detritus, the flow through the outlet is simply set to zero. A partial blockage can be represented by specifying the degree of blockage (proportion of pre-blockage flow) that can pass through the outlet and can be specified (Q_o), thus:

$$Q_o = B_\% Q_{open} \qquad (6.7)$$

where $B_\%$ = percentage of unblocked flow that can pass through outlet, Q_{open} = total unblocked flow.
  The flow from the lower edge of a sloping roof to the collection gutter can be calculated assuming weir flow, thus:

$$Q_o = 1.5 L y_{r_i}^{1.5} \qquad (6.8)$$

where L = length of roof discharging to gutter.

6.2.3 Model application

*Roof flow conditions – impact of rainfall intensity and roof
construction*

Figure 6.8 shows the runoff (into the connected gutter) from a small industrial roof (24.5 m deep, 50 m long, 0.095 slope) as simulated using the
ROOFNET simulation model, which incorporates the solution techniques
detailed in Section 6.2.1. As shown, increasing the rainfall loading from a
current extreme event, i.e. 1 in 100 year – 15 min duration – 404 mm/hr
peak (Institute of Hydrology 2005) to a possible future extreme rainfall event
data, i.e. 566 mm/hr peak an uplift of 1.4 (Evans *et al.* 2004), results in a
significant increase in the roof runoff to the connected gutter. Figure 6.8 also
indicates that replacing the standard roof surface with a green roof effectively
mitigates against such an increase. It is also interesting to note that, once the
green roof is totally saturated (~370 s), it can no longer infiltrate any rainfall,
and it behaves exactly as a standard roof surface, i.e. what lands on the roof
forms runoff and enters the gutter.

6.3 Conventional systems

The flow conditions in the collection gutters within conventional rainwater
drainage systems are normally independent of the conditions in the connected
system pipework. However, as detailed in the following sections, there are
instances when this is not the case.

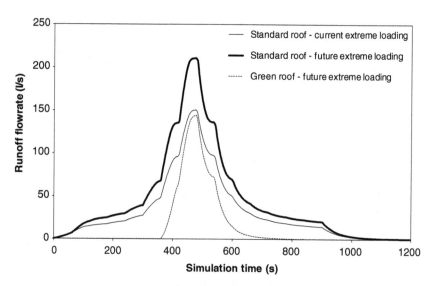

Figure 6.8 Impact of roof construction on runoff to gutters.

6.3.1 Numerical approach – collection gutters

The free surface flows in rainwater gutters may also be simulated via the MoC methodology discussed for partially filled pipeflow. The modification to the momentum and continuity equations necessary to include lateral inflow has already been developed and applies to free surface gutter flow and to land drains. In the case of rain water gutters the gutter slopes are extremely shallow so that subcritical flow would be the norm. The St Venant C^+ and C^- characteristic equations, including a distributed lateral inflow along the length of a drain or rain gutter, were developed earlier as

$$u_1^P = \left[h_1^R + \frac{g}{c}h_2^R - g(S-S_0) + q(u-c)/A\ \Delta t \right] - \frac{g}{c}h_2^P = K1 - K2h_2^P = 0 \quad (6.9)$$

provided that

$$x^P - x^R = \left[u_1^R + c^R \right]\Delta t$$

$$u_1^P = \left[u_1^S - \frac{g}{c}h_2^S - g(S-S_0) + q(u+c)/A\ \Delta t \right] + \frac{g}{c}h_2^P = K3 + K4h_2^P = 0 \quad (6.10)$$

provided that

$$x^P - x^S = \left[u_1^S - c^S \right]\Delta t$$

In common with other applications of MoC the simulation requires initial flowrate and depth values at each node at time zero, based on a 'trickle' flow, and schemes to determine these values utilise the gradually varied flow integration, equation 3.82 (French 1985). However these initial values assume that the gutter is continuously supplied from upstream with a lateral inflow, whereas in the gutter case the upstream boundary would be a dead end. The simplest solution is to use the gradually varied flow approach to determine flowrate and depth and then at the start of the simulation set the upstream boundary to zero flow generating a depth depression wave. Following several pipe periods a stable initial condition is achieved and the simulation proper may commence. Figure 6.9 illustrates this effect; a steady profile is established in around 100 seconds, or with a wave speed of approximately 1 m/s (Figure 2.3), 5 pipe periods. In practice this initial period may be 'hidden' from the simulation by running for a period before initiating the reported simulation; however this is merely cosmetic.

6.3.2 Numerical approach – system pipework

Under free surface gutter conditions, downpipe flows can be assumed to be free surface, and can thus be routed to ground level assuming annular flow

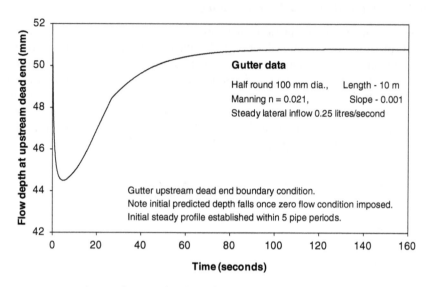

Figure 6.9 Steady initial gutter depth profile established following imposition of a zero upstream inflow boundary condition.

within downpipes (see Section 3.4.5). When the conditions at the gutter outlets in conventional systems incorporating wide gutters become submerged, the flow in the connected downpipes can change to become full bore. An MoC technique can then be used to solve the continuity and momentum equations of one-dimensional, unsteady full bore flow. With reference to Figure 6.10, the St Venant C⁺ and C⁻ characteristic equations for full bore flow were developed earlier as:

$$u^P = \left[u^R + \frac{p^R}{\rho c} - \frac{gsin\alpha + fu|u|}{2m} \right] - \frac{p^P}{\rho c} = K1 - K2p^P = 0 \qquad (6.11)$$

provided that

$$x^P - x^R = \left[u^R + c^R \right] \Delta t$$

$$u^P = \left[u^S - \frac{p^S}{\rho c} - \frac{gsin\alpha + fu|u|}{2m} \right] + \frac{p^P}{\rho c} = K3 + K4p^P = 0 \qquad (6.12)$$

provided that

$$x^P - x^S = \left[u^S - c^S \right] \Delta t$$

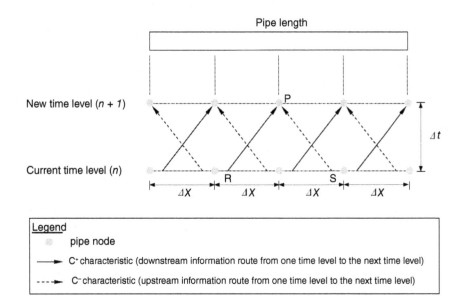

Figure 6.10 Schematic representation of the Method of Characteristics applied to full bore flow conditions.

By utilising the two-step approach detailed above, it should be appreciated that the process of 'priming', whereby downpipe flows gradually become full bore, is not accurately simulated. This issue is discussed in more detail in Section 6.4.3.

In common with other applications of the MoC, the simulation requires initial flowrate and depth values at each node at time zero. For full bore flows this is not an issue as initial conditions may be zero flow and atmospheric pressure.

6.3.3 Boundary conditions

Typical outlets within gutters in conventional drainage systems are essentially 'holes' in the gutters, and when normally freely discharging, the normal subcritical flow will exit the outlet at Critical depth; hence the boundary is adequately described by the condition introduced in Section 3.4.2 for freely discharging pipe flow, which may be recast as:

$$y_o = y_c = \sqrt[3]{\frac{Q_o^{\,2}}{gD_o^{\,2}}} \tag{6.13}$$

where y_o = outlet depth, y_c = Critical depth, Q_o = outlet depth, D_o = outlet diameter.

Where 'wide' gutters are installed (i.e. where gutter sole width >> gutter depth), it is more appropriate to treat the outlets as flat roof outlets, with the boundary being described by solving equation 6.5 with the available characteristics (C^+ from left hand side and C^- from right hand side). If the flow in the downpipe becomes full bore, the conditions within the gutter are inextricably linked to those within the connected downpipe, and the boundary condition is as for outlets on flat roofs, i.e. the available C^- characteristic equation at the entry to the downpipe is solved with an expression linking gutter flow depth to pipe pressure (equation 6.6).

As for outlets on flat roofs, partially blocked gutter outlets can be simulated by representing the degree of blockage (proportion of pre-blockage flow) that can pass into the downpipe, with equation 6.7 being solved with the available characteristics (C^+ from left hand side and C^- from right hand side). Similarly, fully closed outlets can be simulated by setting the outlet flow to zero and solving the available characteristics for gutter depth (C^+ from left hand side and C^- from right hand side).

At the each end of a collection gutter, there is no flow, and hence the velocities at the end nodes are set to zero and the available characteristic is solved for gutter depth.

Gutter overtopping can be simulated using sharp crested weir theory, as developed previously for the discharge from a sloping roof to a collection gutter (see equation 6.8).

As the system pipework is connected to the gutters via the gutter outlets, the outlet boundary conditions developed above are common to the system pipework, i.e. the outflow from the gutters is the inflow to the downpipes. An additional boundary condition is also required for downpipe exiting full bore flow. Here, the C^+ characteristic available at the system exit can be solved with an expression relating exit head loss to flow. Inclusion of a hydrostatic exit pressure term also accounts for the possibility of discharge to a submerged or sealed manhole, thus:

$$p_{j+1} = \max\left(p_{atm}, p_h\right) + \rho g \left(k \frac{V^2}{2g} \right) \qquad (6.14)$$

where p_{j+1} = pressure at exit node, p_{atm} = atmospheric pressure, p_h = hydrostatic pressure due to submerged or sealed manhole, ρ = flow density, k = exit loss coefficient, V = flow velocity.

6.3.4 Model application

The predicted data presented below comes from the FM4GUTT and ROOFNET simulation models which uses the MoC solution technique detailed in Section 6.4.1.

Gutter conditions – impact of rainfall intensity

Figure 6.11 shows the conditions predicted by the FM4GUTT simulation model within a 10 m long, trapezoidal gutter (base width = 100 mm, depth = 200 mm, side slope = 45°, longitudinal slope = 0.001, Manning's n = 0.009) when subjected to time varying inflows (1 l/s for 200 s, 3 l/s for 100 s, 1 l/s for 100 s). As mentioned in the discussion of interpolation techniques, the sub-critical drawdown profile that exists upstream of the Critical depth discharge presents problems in that small rounding errors inherent in a linear interpolation may lead to profile collapse. The FM4GUTT simulation model uses the more accurate Everett and Newton-Gregory interpolation techniques, which maintains the profile as the lateral inflow changes relatively slowly. For these stability reasons it is advisable to run FM4GUTT for a low intensity initial steady lateral inflow period to allow the gutter depths to stabilise, as illustrated in Figure 6.11b.

Gutter outlet conditions – outlet conditions in a standard gutter

Figure 6.12 shows a comparison between the measured and predicted (ROOFNET) outlet depths and overtopping rates for a typical trapezoidal gutter used in small-scale developments (80 mm sole width, 112 mm top width, 60 mm height) and connected to a 64 mm diameter downpipe, when the total inflow to the gutter was approximately 5.9 l/s. As shown, the predicted outlet depth compares well to the measured data, the only significant

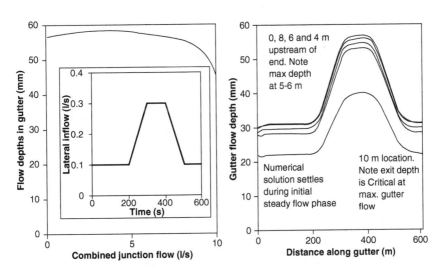

Figure 6.11 Gutter flow depths in a trapezoidal gutter in response to a varying rainfall intensity.

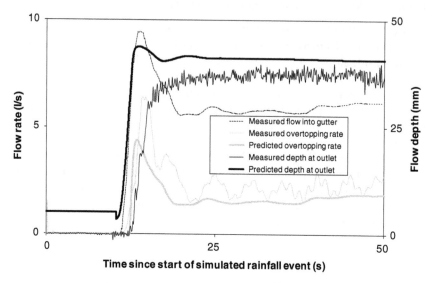

Figure 6.12 Measured and predicted conditions for a standard gutter.

difference being due to the 'fictitious' initial baseflow necessary to commence the simulation (0.1 l/s). The predicted gutter overtopping rates also compare well to the measured data, the only significant difference being due to the highly turbulent initial 'surge', as the gutter first overtops. As ROOFNET simulates one-dimensional flow conditions (along the gutter length), it is not possible to predict variations in flow conditions across the gutter, let alone this type of turbulent gutter conditions. The minor discrepancies shown in Figure 6.12 between the 'cyclical' measured overtopping rates and the 'steadier' predicted rates are again considered to be as a result of turbulence.

Gutter outlet conditions – outlet conditions in a wide gutter

Figure 6.13 shows a comparison between the measured and predicted (ROOFNET) outlet depths for a typical wide gutter (600 mm sole width, 300 mm deep, with a custom-made outlet set flush to the gutter sole), when the inflow to the gutter was reduced in steps from an initial 26 l/s. As shown, the predicted data compares well to the measured data at all gutter inflow rates.

Gutter overtopping

Figure 6.14 illustrates a typical application of the ROOFNET model to the analysis of a conventional rainwater drainage system. This figure shows the simulation results obtained when a 75 m length of 150 mm diameter half round gutter, connected to four 100 mm diameter downpipes, is subjected to rainfall events with intensities of 75 and 100 mm/h. As shown, during the

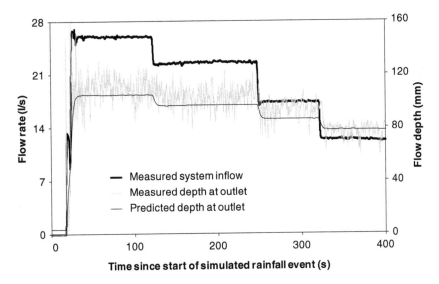

Figure 6.13 Measured and predicted conditions for a wide gutter.

less intense event, the system performs well and no overtopping is predicted. However, during the more intense event Figure 6.14 illustrates that the depths will exceed the gutter top level at certain points, resulting in gutter overtopping and effective system failure.

6.4 Siphonic systems

When operating as designed (i.e. under full bore conditions), the flow conditions in the collection gutters within a siphonic rainwater drainage system are inextricably linked to the conditions in the connected system pipework. Under all other conditions, the situation is more complex.

6.4.1 Numerical approach – collection gutters

Collection gutters within siphonic rainwater drainage systems are dealt with exactly as detailed in Section 6.3.1 for conventional systems.

6.4.2 Numerical approach – system pipework

Given the range of different conditions that can occur within a siphonic system, any numerical scheme must be able to simulate both free surface and full bore flow conditions, as well as the intermediate phase where both flow conditions may exist in the pipe network simultaneously.

The unsteady flow conditions experienced within the siphonic rainwater drainage network may be described by the appropriate full bore and partially

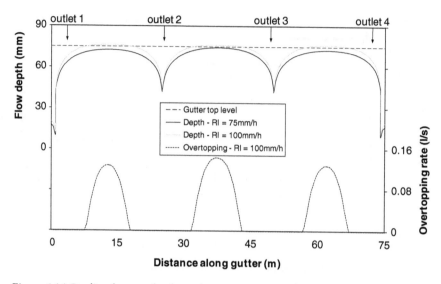

Figure 6.14 Predicted gutter depths and overtopping rates for a 75 m section of Gutter B connected to four 110 mm diameter downpipes (roof area = 650 m²).

filled pipeflow versions of the St Venant equations as defined previously. However while the Method of Characteristics solutions are appropriate for the separate full bore and partially filled cases, difficulties arise when it becomes necessary to track multiple hydraulic jumps through the network, as is the case as the flow moves to siphonic full bore. The wide-ranging movement of hydraulic jumps is an essential element of system priming, and the MacCormack method developed previously can be employed to simulate initial free surface flow conditions, whilst retaining the MoC technique for the simulation of full bore flow conditions. The transition between the two flow regimes can be dealt with using the Preissmann slot technique developed in Chapter 3.

It should be noted that, as the MacCormack method is not readily applicable to vertical pipework, the free surface flow conditions in the downpipe must be assumed to be annular, whilst the filling of this pipework (just before the system primes) can be simulated using a volumetric-based technique.

As the vertical downpipe within a siphonic system starts to fill, the system will start to depressurise, thus rendering the MacCormack method unsuitable. At this point it is necessary to switch over to the classical Method of Characteristics solution technique of the governing equations of full bore flow, as developed previously for full bore flow conditions within conventional systems. When the system starts to depressurise, a suitable model can check for regions where free surface conditions exist; it is these regions that correspond to the trapped air pockets observed during the experimental work. Once the system is primed, these air pockets can be then tracked as they leave the system, their velocities being set equal to that of the local water flows and their volumes/pressures being dependent on the universal gas laws and local water pressure conditions. Depending on system layout, certain types of air pockets may become mixed with the water flow before exiting the system, hence forming 'bubbly flow'.

The following additional points concerning the computational principles underlying a suitable numerical model are worth noting:

- Once a system flows full bore, it can be assumed that any flow entering the system is a homogeneous air/water mixture; the air content, and hence wave celerity, being a function of the gutter water depth.
- As there is no satisfactory, theoretical method of simulating the type of pulsing flow conditions that occur downstream of three pipe junctions under certain circumstances, such conditions can be represented by assuming full bore flow conditions with a high air content (based on the relative flow rates converging at the junction). To simulate the additional head losses incurred due to the pulsing nature of the flow, an increased pipe roughness value can be used downstream of any such junctions.
- As well as tracking the progress of the full bore front in the vertical downpipe, it is also necessary to track the progress of any free surface fronts as these can lead to the draining of individual system branches or complete systems as rainfall intensities decrease.
- As with all explicit finite difference schemes, the time step must be chosen to satisfy the Courant Criterion.

The numerical approach outlined above is incorporated into the SIPHONET simulation model and has been proven to be accurate. However, the use of two different computational techniques can result in numerical instabilities, particularly at the application interfaces, and such problems are often exacerbated by the need to keep track of multiple hydraulic jumps and air pockets within complex siphonic rainwater drainage systems. Consequently, the SIPHONET model may not always be sufficiently robust for everyday design and diagnostic purposes. To overcome such computational difficulties, it is necessary to simplify certain aspects of the numerical modelling approach applied to system pipework. One such approach is followed in the ROOFNET model, which can simulate conditions within both conventional and siphonic systems. Whilst it deals with flows conditions on roof surfaces and in gutters in an identical fashion to SIPHONET, it employs a two-step approach which is applied to the simulation of flow conditions within siphonic pipework. If the flows entering the system are insufficient to induce full bore conditions, the flow is simply routed through the pipework. However, if the flows entering the system are sufficiently high, full bore conditions occur and a full method of characteristics solution is then used to simulate flow conditions. Although this approach simplifies the initial free surface pipe flow conditions, it does accurately simulate full bore flows within the pipework, as well as the roof and gutter flow conditions. Furthermore, the model is numerically stable, and hence generally applicable. As such, the ROOFNET model can help inform designers as to the likelihood of system priming system operation, as well as identifying when and where overtopping and flooding events may occur.

6.4.3 Boundary conditions

In the case of gutter flow, the boundary conditions are similar to those previously detailed for conventional systems. The only significant differences are in the weir coefficients used and in the additional head loss term in the full bore gutter outlet boundary condition to account for losses in the short vertical 'tailpipe' and connected bend (see Figure 6.7).

As with conventional systems, the gutter outlet boundary conditions detailed previously are common to the system pipework, i.e. the outflow from the gutters is the inflow to the downpipes.

Prior to system priming, when conditions within a siphonic system are free surface, the necessary boundary conditions are identical to those already developed for free surface pipe flow (e.g. pipe junctions) or for flow in conventional rainwater systems (e.g. annular flow, freely discharging system exit). Similarly, once a siphonic system has primed, the system exit is treated in an identical manner to the exit from a conventional downpipe running full bore; i.e. the C^+ characteristics available at the system exit are solved with equation 6.14, relating exit head loss to flow. However, under full bore flow conditions, boundary conditions are required to represent pipe junctions. These junctions include straight through junctions (connecting different diameter pipes), two and three pipe junctions and junctions between the horizontal pipe work and the vertical downpipe. Irrespective of the type of junction, the available C^+ characteristic(s) in the upstream pipe(s) is solved with the single C^- characteristic in the downstream pipe, and an expression relating the junction pressure loss to the junction through flow. Typically this expression is of the form:

$$P_L = \sum_{i=1}^{n} \rho_i g k_i \frac{V_i^2}{2g} \qquad (6.15)$$

where p_L = pressure loss through junction of n pipes, ρ = flow density, k = exit loss coefficient, V = flow velocity.

This boundary condition can also account for the confluence of flows with different air contents, and the possibility that one of the junction branches reverts back to free surface flow conditions, i.e. drains due to insufficient inflow.

6.4.4 Model applications

Except where noted, the model data presented below refers to the siphonic system shown in Figure 6.15, which is part of the Heriot-Watt experimental drainage facility, and uses the SIPHONET model.

Design criteria rainfall

Figure 6.16 shows the measured and simulated gutter depths and system pressures as the system primes under the design rainfall criteria. As shown, the model data accurately predicts the gradual depressurisation of the system as the vertical downpipe starts to fill, as well as the re-pressurisation that occurs when the main air pocket leaves the system. The only significant discrepancy in these figures are the oscillations that occur when the system starts to depressurise, and the model switches from using the MacCormack solution technique to the Method of Characteristics. These may be considered temporary 'adjustment' errors; a fact borne out by the accuracy of the model immediately prior to and immediately after this switch has occurred.

Rainfall events below the design condition

Figure 6.17 shows the measured and simulated gutter depths and system pressures when the system loading is below the design criteria, resulting in

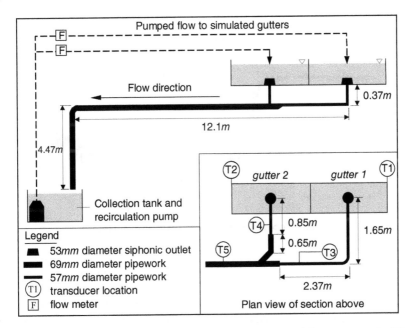

Figure 6.15 Schematic view of the Heriot-Watt siphonic roof drainage test rig. The capacity of the system is the same as 75 mm/hour falling on a 665 m² roof – all drained via a 69 mm diameter pipe.

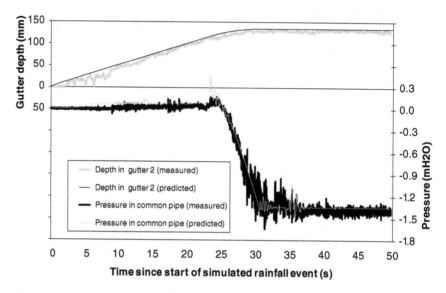

Figure 6.16 Measured and predicted conditions for design criteria rainfall.

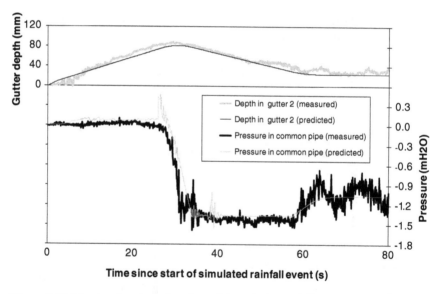

Figure 6.17 Measured and predicted conditions for a rainfall event below the design criteria (gutter 1 inflow = 5.10 l/s; gutter 2 inflow = 6.10 l/s).

insufficient gutter inflows and hence breaking of full siphonic conditions (at about 60 seconds). As shown the model accurately simulates the decrease in gutter depth that occurs as a result of insufficient gutter inflows, as well as predicting the increased air entrainment and decreased flow densities that result in an increase in the steady state system operating pressures.

Siphonic roof dynamic balancing

Figure 6.18 shows the measured and the predicted depths in gutter 1 and pressures in the collector pipe downstream of the junction in response to a simulated rainfall event. In addition to showing the formation of siphonic conditions (0 s to 32 s) and a period of steady siphonic action (32 s to 62 s), this data also illustrates the rise in system pressures that is transmitted throughout the system when the depth in the gutter drops below that necessary for full bore flow, hence allowing air to enter the system and break the siphon (at approximately 62 s). Clearly, if the inflow to gutter 1 was not restarted (at approximately 82 s), siphonic action would not have been re-established and the continuing inflow to gutter 2 would have led to overtopping of that gutter, and hence system failure.

Blockage of an outlet

Under fully operational conditions the gutter outlets admit high rates of inflow to the network. Any inadvertent reduction in this flow will generate negative pressure transients. The generation of large negative pressure transients within a system already subjected to sub-atmospheric line pressure may be sufficient to cause pipe section implosion with consequent flooding of the serviced building.

Figure 6.19 shows the predicted pressure in the pipe network in response to an instantaneous and total blockage to the outlet in gutter 2. This results in the generation of a negative transient; as the simulated blockage was instantaneous and total, the pressure drop associated with this transient is

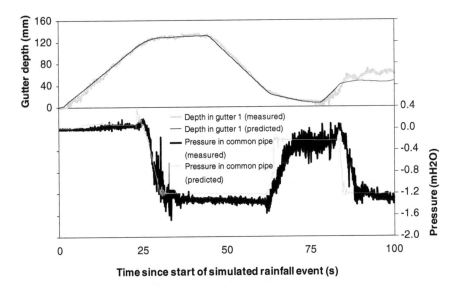

Figure 6.18 Measured and predicted gutter depths and system pressures from the MacCormack/MoC hybrid simulation.

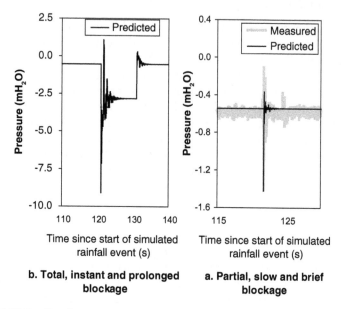

Figure 6.19 Predicted pressure surge generated by an instantaneous gutter outlet blockage.

equal to the Joukowsky pressure drop. Pressure transients of this magnitude would almost certainly result in the implosion of the type of pipework commonly employed in siphonic rainwater drainage systems.

Submergence of the system exit

To ensure the efficient operation of a siphonic roof drainage system, it is essential that full bore flow conditions are broken before any connection to the surface water sewer network. If not, the flows within the siphonic system and the sewer network may interact, leading to unpredictable conditions and potential problems. Breaking of full bore flow conditions can only be guaranteed by ensuring that the flow exits the siphonic system above the highest water level in the surface water sewer. However, as surface water sewers are normally designed to a lower level of risk than roof drainage systems, it is clear that a rainfall event which causes the priming of a siphonic roof drainage system may also cause surcharging of the downstream surface water sewer; a scenario that could lead to the type of flow interactions discussed above. Figure 6.20 shows the predicted system characteristics when the vertical downpipe becomes submerged. As shown, submergence of the system exit results in a decrease in the available driving head, an increase in system pressures and a resulting decrease in system flow rates. Such events could lead to an increase in the time required to prime a siphonic system, and

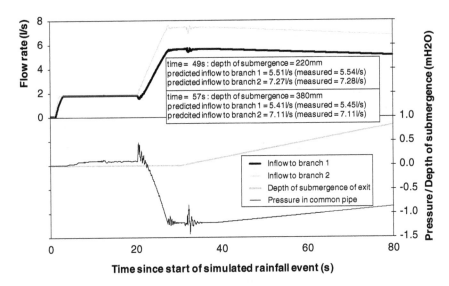

Figure 6.20 Predicted conditions for design criteria rainfall event, with gradually submerging system exit.

will certainly reduce the total capacity of the system; either of these scenarios could lead to failure of the system by gutter overtopping.

Robust system modelling

Figure 6.21 shows the simulation results obtained from the ROOFNET model, using the simplified modelling technique, where an extreme event rapidly fills a siphonic system incorporating three outlets. After operating normally for a short period, the system capacity can be seen to decrease between 20 s and 30 s. This is due to the submergence of the system ground level sewer exit, which results in a lower effective driving head and may be considered analogous to the surcharging of the downstream sewer connection. Finally, at approximately 40 seconds, the effect of blocking an outlet is illustrated. Although the flow through outlet number 1 reduces to zero, it can be seen that the flows through the remaining outlets actually increase, a result of the change in the balance between system losses and system driving head.

6.4.5 Site monitoring

As part of the research underpinning the development of numerical simulation models, the conditions within a number of siphonic systems installed on the buildings housing the National Archive of Scotland (Edinburgh) have

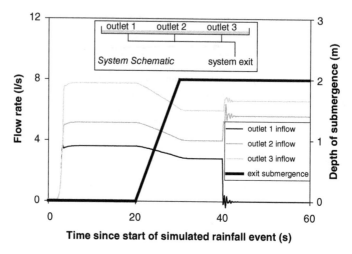

Figure 6.21 Predicted flow rates for a siphonic system experiencing exit
submergence and outlet blockage.

been monitored since June 2000 (Arthur *et al.* 2005). Figure 6.22 illustrates
the measured gutter depths and system pressures alongside the recorded
rainfall intensity, for the most extreme event to date. This event had a maximum rainfall intensity of 105 mm/h, which equates to a return period of 67
years (31 mm rainfall over 94 min), and appeared to result in continuous
siphonic action for a period of approximately 500 s.

In addition to validation data, the long-term monitoring undertaken at the
National Archive of Scotland, and other sites since, has also highlighted the
key role that maintenance plays in the operational characteristics of siphonic
rainwater drainage systems. As illustrated in Figure 6.23, even when systems
are correctly fitted with leafguards, inadequate maintenance can lead to outlet blockage and impaired system performance.

6.5 Summary

This chapter has detailed the application of various numerical approaches
to the simulation of flows within both conventional and siphonic rainwater
drainage systems.

The three key components of any rainwater drainage system (roof
surface, collection gutters and system pipework), and their impact of prevailing flow conditions, have been described. Conditions on roof surfaces
are shown to be strongly 2D and typified by low flow depths in relation to
flow widths, whilst those in collection gutters can be approximated as 1D
and can have flow depths comparable to flow widths. Similarly, it is shown
that the conditions within system pipework can also be approximated as

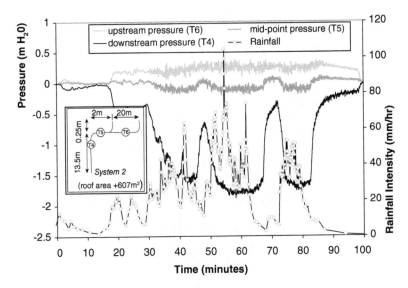

Figure 6.22 Gutter depth and pipe network pressures during an extreme rainfall
event at the National Archive of Scotland test site installation.

1D. However, whilst conditions on roof surfaces and in collection gutters
are not dependent on the type of system used, the flow conditions within
system pipework is shown to be strongly influenced by system design; flow
conditions within conventional downpipes is annular, whilst those within
siphonic systems can vary tremendously, from annular through to pulsing
two phase and full bore.

Current design methods are discussed, and it is concluded that the current
dependency on empirical relationships and/or simplifying assumptions limits the applicability of such methods to real-world operational performance
issues.

The remainder of the chapter focuses on the development and application
of suitable numerical simulation models for rainwater drainage systems. A
key element of the simulation models presented is to ensure that the numerical techniques used are tailored to the characteristics of specific system
components. Hence, a simple finite difference solution of the kinematic wave
approximation to the full governing St Venant equations is proposed for the
2D flow on roof surfaces, as opposed to more computationally demanding
fully dynamic approaches. Key boundary conditions are discussed, and the
developed model is shown to be capable of simulating the impact of different
roof designs (e.g. green roofs) as well as that of variable rainfall and wind
loading. Similarly, an MoC-based solution of the full governing equations
is introduced to simulate the predominantly 1D free surface flow conditions
within roof gutters and in system pipework running full bore. Free surface

Figure 6.23 Blockage of a siphonic roof outlet.

flow conditions within conventional drainage system pipework is shown to be adequately described using the assumption of annular flow, whilst a hybrid MacCormack/MoC modelling technique is introduced to simulate the more complex conditions typically found within siphonic pipework. The numerical 'robustness' of hybrid models is discussed, with respect to the simulation of conditions within siphonic systems, and a simplified approach is proposed for everyday design and diagnostic use. Again, the importance of suitable boundary conditions is emphasised. The developed models are shown to be capable of accurately simulating conditions within all types of rainwater drainage systems, under a variety of different loading regimes.

7 Design applications

7.1 Introduction

Water conservation is seen as an inherent prerequisite to the sustainable provision of utility services within the building envelope; however concerns have been raised as water usage appears to be reducing, that there will not be enough water to keep the drains clean. The concept that 'dry drains' could be a probable consequence of water conservation arose in the US where 'concerned groups' argue that further reductions in w.c. flush volume should be resisted as current water conservation measures reduce the throughflow in the drains to the extent that 'drainline carry' becomes impossible. This question is dealt with in this chapter together with the need for simulation and in-depth analysis of drain line carry in the context of water conservation strategies. The implications of reduced flush operation on drainage network design will also be discussed, using two case studies: one which looks at the implications of increasing collection drain size as a result of a quirk in legislative prudency (changes to building regulations in England and Wales) and the other which looks at the implications of allowing even further reductions in flush volume in building regulations with the goal of improving sustainability and reducing carbon emissions in Scotland. The case studies make extensive use of the simulation program DRAINET to evaluate optimum operating condition. Drain diameter and gradient choice are discussed both within the horizontal branches of the building drainage network and also in the below-ground connections to the sewer network. The importance of wave attenuation, as discussed in Chapter 4 will be reiterated here, and an evaluation of the link between water usage and carbon emissions will be made for the second case study.

The case studies deal with solid transport and the conveyance of waste away from habitable space, in partially filled pipes, to the main sewer. This field of study was chosen over roof drainage since it is the most contentious, and one in which there are very strong views on all sides of the debate. It has also been chosen since it is directly related to the supply of potable water and therefore one in which the greatest impact on water conservation can be achieved. The case studies therefore show the power of free surface wave and solid transport modelling and the effectiveness of this analysis method in influencing policy makers and the legislature.

7.2 Background

The UK Water Supply (Water Fittings) Regulations 1999 recognized the water conservation imperative by introducing new levels of w.c. flush volume and the re-introduction of dual flush w.c. operation. While overall water consumption is a concern, and while water leakage from the supply network is rightly considered a priority area, w.c. water usage is still the single greatest source of potable water use within both domestic and commercial buildings. In the UK w.c. usage consumes 35% of the domestic potable water supply while combined w.c. and urinal operation in commercial buildings approaches 60% of the potable supply, and this level of usage is replicated in other developed countries.

The 7.5 litre flush volume introduced from 1988 has been reduced to a maximum of 6 l with dual flush provision set at a maximum of two-thirds of this level. In the main these volumes are reducing further with 4 1/2 l flushes not unheard of in many jurisdictions. Necessarily, such regulations cannot be retrospective and hence the majority of the w.c.'s installed in the UK will still be operated on the previous pre-1988 standard of 9 l or above – Scotland accepted a 13 l standard for many years. With 45 million w.c.'s installed in the UK and a turnover rate of 2 million per year the current levels of water use will take time to reduce. Urinal water usage is being reduced by the progressive introduction of user-sensing flushing devices to replace the automatic 20 minute cycle flushing that was the UK standard approach for decades, and waterless urinals are much more commonplace than before; however the norm is still a water flushing urinal. Reductions in water usage will therefore be progressive; however the 1999 Water Regulations were an essential first and significant step in the UK.

Similarly, domestic and commercial washing machine and dishwasher cycles are progressively using lower quantities of water, while the slow penetration of 'grey' water reuse and rainwater storage schemes offer possibilities for future reductions in demand.

Implementing water conservation is difficult if it is reliant upon achieving changes in public attitudes. More productive is an approach based on reductions in appliance usage that do not require any change in user habits. This approach delivers water conservation without the need for swings in public perception and should form the basis for future developments. However this places a greater burden on the appliance manufacturer, whether of water closets, urinal flushing control mechanisms, washing machines or dishwashers to develop units that meet the user's requirements and aspirations. Failures and 'difficulties' attributable to lower water usage must be minimized.

While water conservation is seen as a 'good thing', it must also be understood that there is an economic and carbon reduction rationale for water saving. The level of infrastructure provision that will be needed to support the housing provision required, particularly in already heavily populated regions of the UK, is considerable. Water stress in these areas is inevitable. Water conservation has an important part to play in reducing the levels of new investment and land usage that would be required to supply the higher levels of water storage and usage – assuming that the water resource existed. Similarly reductions in water use imply further economies as the existing drainage infrastructure could handle increased occupancy following change of use – a possibility in redeveloped 'brown' site building programmes.

It may be argued that the advantages of water conservation will be offset or wholly negated if the outcome for the consumer, either in the domestic or commercial sense, is a rise in maintenance charges brought about by inefficient drain cleansing due to reduced throughflows. Thus water conservation addressed through reduced w.c. flush volumes, reduced machine cycle water consumption or a move to reduced-flow showers and user-sensing urinal flushing must be accompanied by an assessment of drain sizing that will yield comparable performance at the reduced flows likely to be encountered in the future. The downward trends in water usage over the past 20 years in all these category areas are well documented, most recently in the Scottish domestic water consumption survey of 1999. Reductions of 50% or more have been achieved in the w.c. flush volumes of appliances now available to the consumer. A comparison of w.c performance from the early 1980s to the present day demonstrates that improved or at worst comparable performance may be expected at much reduced flush volumes due to improved water efficient design. However no comparable changes to drainage design criteria have been introduced that would guarantee comparable system performance. To some extent the regulatory situation in the UK is complicated by the separation of the relevant regulatory committees. The Water Regulations Advisory Committee, responsible for the 1999 Water Regulations and hence the appliance water usage levels, operates under the auspices of Defra. The Building Regulations Advisory Committee, responsible for the definition of drainage network sizing through Part H of the Building Regulations, 7 was the ODPM, previously DTLR and now covered by the Department of Communities and Local Government (DCLG).

The way in which a government operates depends largely on how the departments are organised, and many decisions which influence relevant legislation depend on the effectiveness of the bureaucracy. This is true for all governments and legislatures. It is therefore the role of the expert engineer to inform policy at the highest level using the most effective tools available.

As the predominant percentage of the drainage flow within the building envelope will emanate from w.c. discharge it follows that the interaction between w.c. operation and the drainage network performance should form the basis for drain sizing design criteria. Solid deposition leading to increased maintenance costs is clearly a concern as throughflows fall, and hence the definition of the linkage between reduced flush volume and drain sizing becomes essential.

7.3 Dry Drains 'myth or reality'

In March 2009 a Dry Drains Forum was hosted at the ISH exhibition and speakers presented arguments for and against the concept (Swaffield 2009a). The US contribution harked back to the 1992 Energy Bill that introduced 6 litre w.c. flush volume without consultation and was concerned that any move towards 4 litres would be achieved in a similar fashion. The UK and Australian contributions emphasised the role of climate change and our ability to design for reduced flush volume – quite different approaches. The meeting concluded with no real consensus; however it was noted that much relevant research had already been done and the power of simulation tools such as DRAINET should be used more to address such wide-ranging questions.

Some important issues that were raised focused on careful system and appliance design which can minimise the probability of deposition while flow booster solutions should also be recognised. Minimum transport distance to the first joining flow junction is an essential parameter and has design implications. While accurate data is not available, experience suggests that in the majority of cases the distance to a flow confluence will be less than 5 metres, the frequency of w.c. to junction distances following the pattern suggested in Figure 7.1. Developing the actual form of Figure 7.1 would be a priority research objective as the mean distance to a flow confluence is a major determinant of drainline carry. The importance of the confluence of flows will feature in the case studies presented later in this chapter.

Similarly junction design, as discussed in Chapter 3, becomes a major issue as the hydraulic jumps upstream of a junction of two or more flows present an impediment to solid transport leading to deposition. Swept entry junctions should be used and top entry 90° entries banned. Flow boosting is also of interest; 1980s research concentrated on two alternate designs: the traditional tipping tank used in the UK from at least the 1860s where an eccentrically pivoted tank tips a large water volume into the drain at periodic intervals, and the siphon tank, accepted by Stockholm in 1989 as a design solution to the installation of 3 litre w.c.'s in city apartment blocks – a 21 litre siphon tank in the basement intercepted w.c. flushes and delivered its contents to the drain in one surge.

The reality of whether dry drains will become a major factor in drainage system design is that it appears unlikely, given that overall water consumption is still on an upward trajectory (Waterwise 2014). The case studies given below represent two 'real life' questions posed in relation to whether w.c. volumes can be reduced further, thus keeping the momentum of water conservation going.

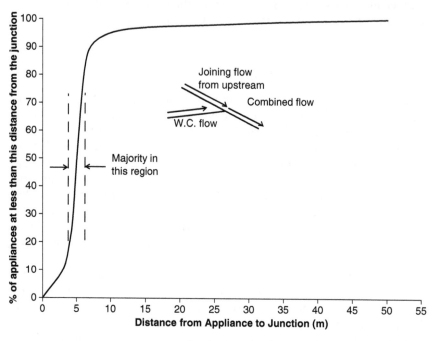

Figure 7.1 Distribution of w.c. to junction distances likely in current practice.

7.4 Case study 1: influencing policy-making on drain sizing

Legislation in the UK features 'adoptable' sewers, namely sewers that become the responsibility of the water companies, hence removing the direct cost of maintenance from the householder. In the past these 'adoptable' sewers have been defined in terms of their diameter, i.e. not less than 150 mm. One strategy to deliver universal adoption would be to ensure that all dwellings were connected to the sewer via 150 mm diameter building drains. However this blanket solution could lead to severe deposition problems due to the low flow expected from single dwellings. Current UK w.c. flush volumes are typically 6/4 litre dual flush; however lower flush appliances are increasingly available – these would exacerbate the drain flow situation for new build properties.

To clarify these issues a simulation was undertaken to determine the likely flow conditions that would be experienced in the dwelling to the lateral drain and in the lateral drain as it collects flow from sequential dwellings. International data on flow loadings were also accessed. The Method of Characteristics simulation including the solid transport model, DRAINET, was used to determine both the solid transport and the flow conditions in a typical dwelling to sewer network. This provided the evidence that led to a modification to the proposed legislation that increased the number of dwellings required to specify an increased drain diameter.

To emphasize the importance of drain diameter a DRAINET simulation was used to compare solid transport within 75 mm, 100 mm and 150 mm diameter pipes immediately downstream of a 6 l flush volume w.c. Figure 7.2 shows the different solid travel distances achieved within these three pipes, which confirm the relationships demonstrated earlier. For the assumed solid, 98 g weight, 38 mm diameter, 1.05 specific gravity, the transport distances varied by up to a factor of 3 as the drain diameter was decreased. This is due to the reduction in wave attenuation and the increase in flow depth surrounding a simulated solid.

The technique used to provide the comparison may also be extended to include solid transport within complex networks subject to a random discharge of appliances and may therefore be used to investigate the building to lateral drain connection sizing issue. In order to simulate the more complex multiple dwelling network, data on appliance usage patterns would be required, and these are still scarce.

Figure 7.3 covers the w.c. and washbasin usage and was used to give confidence to the 'morning rush hour' pattern used in this case study. Three different house types with a range of appliances were simulated, as shown in Figure 7.4. An assumed usage pattern for the morning rush hour was used within DRAINET to determine the dwelling discharge taken to represent the flow profile at the base of a vertical stack serving all the appliances within the dwelling. This usage pattern includes w.c. usage, showers, washbasins, a single bath and dishwasher rinse cycle operation. All dwellings were assumed to be two storey homes with a mixed adult/child population. It was also assumed that the peak usage for each individual dwelling would occur within a 30 minute period.

The assumed layout within each two storey house is shown in Figure 7.5, to include all the appliances to be operated within the simulation of

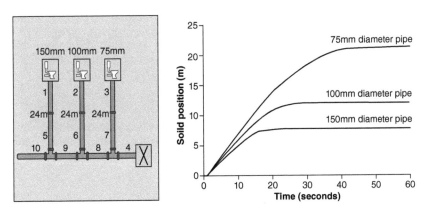

Figure 7.2 Demonstration three pipe network and a comparison of the transport distances achieved in 75 mm, 100 mm and 150 mm diameter pipes by a 6 litre w.c. discharge at a slope of 1/100.

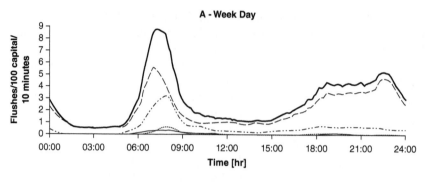

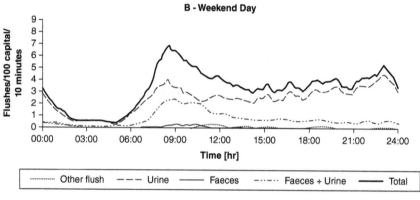

Figure 7.3 Diurnal w.c. usage patterns. (Reproduced from Eran Friedler, David M. Brown and David Butler (1996) 'A study of WC derived sewer solids', *Water Science and Technology*, Vol. 33 No 9 pp. 17–24, with permission from the copyright holders, IWA Publishing.)

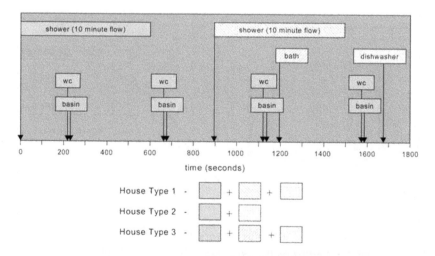

Figure 7.4 Assumed dwelling usage pattern morning 'rush hour'.

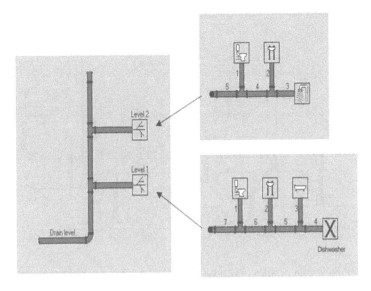

Figure 7.5 Layout of sanitary fittings and drainage.

multi-appliance/multi-house operation. DRAINET was used to generate the flow profiles illustrated for each of three house types in Figure 7.6. The w.c. contributions are clearly identifiable, as are the bath and dishwasher flows. Note that 50% of the w.c. discharges were assumed to contain a solid of 98 g weight, 38 mm diameter and 1.05 specific gravity.

Three house types were used in the broader study – these are character-ized by additional use of a bath and a dishwasher. Based on the accepted 1% level for simultaneous operation of appliances – where the houses with the discharge patterns already developed are assumed to be the equivalent of appliances with a 30 minute operation cycle within a 90 minute observation period – the overall operation to be simulated by DRAINET was randomly determined, as shown in Figure 7.7. It should be stressed that there is only a 1% chance of such heavy simultaneity occurring within any such group of nine houses. A 20 minute test period was chosen such that it would have the greatest combined flow volume from all nine houses, which in this case occurs from 46 minutes to 66 minutes within the 90 minute overall period. Note that, due to this, the test period for some houses starts part-way through their 30 minute operation cycle.

The flow profiles shown in Figure 7.6 were used to 'drive' a simulation for nine houses connected to a sewer. The house connection and the collector drain (sewer) were assumed to be either of 100 mm or 150 mm diameter, at a slope of 1/60. (A 75 mm diameter simulation was also demonstrated; however this was included as a benchmark rather than being a practical installation proposal.) The house connection drains, i.e. from the base of the

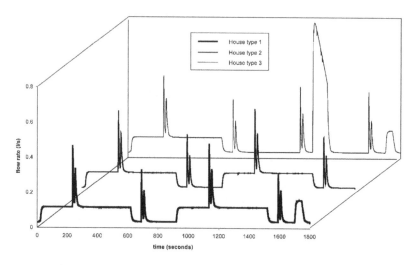

Figure 7.6 Flow rate profiles for house types 1, 2 and 3.

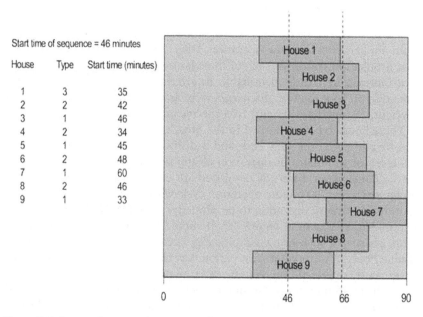

Start time of sequence = 46 minutes

House	Type	Start time (minutes)
1	3	35
2	2	42
3	1	46
4	2	34
5	1	45
6	2	48
7	1	60
8	2	46
9	1	33

Figure 7.7 Assumed pattern for a series of nine dwellings connected to common drainage.

vertical stack to the sewer, were 2 m long, and the individual sewer pipes between houses 10 m long, giving a maximum drain length of 92 m.

The layout of the nine houses and the form of the drainage simulated are illustrated in Figure 7.8. Note that this figure is not to scale. DRAINET was

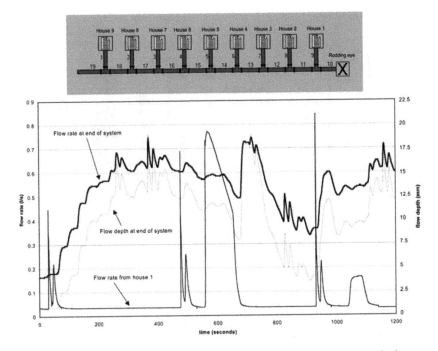

Figure 7.8 Simulated nine dwelling group and cumulative flow rate at end of system for 100 mm diameter pipes. Additionally data presented for the inflow from dwelling 1 and the flow depth at the end of the system.

run to simulate 20 minutes of drain activity. For the 100 mm diameter case, the combined flow rate at the end of the system is also shown in Figure 7.8. Flow rates for 75 mm and 150 mm diameter pipes are not shown, as these are very similar. Note the bath discharge at 680 s into the simulation. This originates from house 1, whose 30 minute operation commenced at 35 minutes into the 90 minute study period. Thus the bath discharges 540 s into the simulation.

Summary data for solid transport for systems constructed from 75 mm, 100 mm and 150 mm diameter drains at both 1/60 and 1/100 slopes are given in Figure 7.9, illustrating the:

- percentage of solids deposited and not transported out of the stack-to-sewer connection drain
- percentage of solids deposited and not transported out of the collector drain
- percentage of solids discharged from the network to a larger sewer – the maximum travel distance required would be 92 m in the network illustrated.

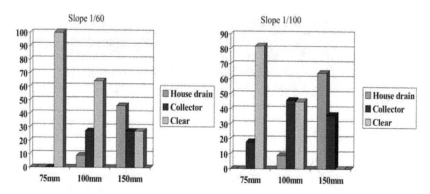

Figure 7.9 Solid transport comparison for overall slopes of 1/60 and 1/100 illustrating the percentage of solids to clear the network and the percentage deposited in either the house to collector section or in the collector drain. The advantage of steeper smaller diameter drains is clear.

The results, shown in Figure 7.9, demonstrate the advantage of the steeper 1/60 slope as well as the improvement in transport performance represented by the smaller diameter drains. Parallel laboratory observations involved noting the travel distance for a 98 g weight, 38 mm diameter, 1.05 specific gravity cylindrical solid in 75, 100 and 150 mm diameter drains at a slope of 1/100 when subjected to the minimum flow- rate necessary to cause motion in a 150 mm diameter drain. The duration of the flow was chosen to represent a 6 l water closet discharge. As expected from the data discussed earlier reducing the drain diameter was advantageous.

The importance of the parameters identified during early empirical work on solid transport within building drainage networks has been confirmed by simulations based on the fundamental equations of continuity and momentum defining unsteady free surface flows. Reductions in w.c. discharge volumes brought about by the growing importance of water conservation will lead inevitably to increased maintenance costs unless consideration is given to matching reductions in drain or horizontal branch diameters.

The simulation techniques discussed and demonstrated lay the basis for such reductions in drain sizing. The simulations are of a form that may be used constructively by code bodies in determining the consequences of changes in prescribed diameters and other system parameters, such as slope and to a lesser extent drain material, as represented by roughness. Similarly the simulations presented are in a form that will allow manufacturers of w.c. and other appliances to predict the effect on system performance of new water conserving appliances intended for the market.

Finally the simulations presented offer a design corroboration tool to system designers concerned about either the impact of water conservation on the operation of their systems or the advantages that may flow from reduced horizontal branch diameters in considering future change of use possibilities for space initially allocated to washrooms.

Water conservation as a central theme of sustainable construction is a long-term consideration for designers. The use of simulation-based design advice is essential if the full benefits of water conservation are not to be reduced through increased maintenance expenditure and system failures.

7.5 Case study 2: implications of reducing w.c. volume

In this case study the impact of reducing flush volumes below the currently accepted level is assessed in the context of operational efficiency and the potential carbon emissions saved.

Approximately one-third of the entire water supply delivered to a home or a commercial building is flushed down the toilet (Environment Agency 2007; Griggs and Shouler 1994). While this is 'wasteful' there are other detrimental effects which should also be considered. The water being flushed into the drain has been processed and cleaned to the highest standards for wholesome drinking water, a process which is costly in both financial and energy terms. The carbon footprint of water supply (Scottish Water 2010) cannot be ignored if carbon emission targets are to be met. The introduction of lower flush volume w.c's, forms part of an effective strategy to make carbon emission savings while at the same time conserving a finite natural resource of great environmental significance.

There are approximately 45 million w.c.'s in the UK as a whole. The total amount of water used per day to flush toilets is estimated at 2000 million litres (waterwise.org). If these figures are disaggregated, pro rata by population, then there are approximately 3.8 million w.c.'s in Scotland. The average number of flushes per w.c. per day is approximately 5 per w.c., and the average flush volume across all w.c.'s in service is 9 litres. Based on figures published by Scottish Water the carbon footprint of producing 1 ML of treated water is 0.15 tonnes of CO_2. It is therefore possible to calculate the CO_2 emissions from w.c. flushing in Scotland. The annual CO_2 emissions from treating water which is flushed back down the drain is approximately 9200 tonnes. Reducing the flush volume of every w.c. in Scotland by 1 litre could contribute to a saving in CO_2 emissions of 920 tonnes.

It is widely accepted that using high quality potable water within some appliances, such as w.c's, is a waste of a resource, especially in countries where water of such a high quality is in rare existence and reasonably inaccessible (UNWAP, and UNESCO 2004, Dolnicar and Schafer 2009). Water consumption in Scotland has increased over the years. Moran *et al.* present figures that suggest an increase of 35% in the supply of water between 1971 and 2001 for unmetered properties (Moran *et al.* 2007). In 2001 the domestic water demand was estimated at 144.03 litres per person per day. This is expected to rise to 150.32 litres per person per day by 2015, an increase of 4.4% over the period (Moran *et al.* 2007). There are factors such as climate, culture and economy that will have an impact on the domestic water consumption in different countries as discussed by Lazarova *et al.*, who also discuss the factors that result in variable per capita consumption, including age, gender, type of domestic appliance and metering arrangements (Lavarova 2001). While some data on Scottish water consumption and appliance usage patterns exists it is still scarce, and further research into this important area is required to more fully understand the Scottish perspective.

The introduction of numerical modelling techniques to building drainage systems has allowed a fresh insight into the workings of systems. The characteristics of solid transportation within a building drain are different from those in a public sewer network. In general this difference is one of scale; inside a building the processes occur quickly, within seconds or minutes, while in a public sewer transportation time can be considered in hours and days. Additionally, the speed with which processes occurs within a building poses a modelling challenge which can be overcome by the introduction of a Method of Characteristics–based numerical model. Modelling using numerical techniques allows new system configurations and appliances to be assessed as long as appropriate boundary conditions can be ascertained or assumed.

7.5.1 Practical aspects of lowering flush volume

The introduction of lower flush volume w.c.'s offers the opportunity to reduce both water usage and the carbon footprint of the water supply network. While this is undoubtedly true there are practical aspects to the process which cannot be ignored. If consumers are encouraged to use different appliances there are other considerations such as cost and availability, the effect on existing drain and sewer pipe operation and identification of lower flush volume appliances which need to be considered.

7.5.2 Cost

The most common w.c.'s currently available are push button, dual flush with flush volumes of 6 and 4 litres, but there are also combinations of 6 and 3, 5 and 3, 4.5 and 3, 4 and 2.6, and 3.5 and 2 litres (BMA Trends 2010). Single flush volumes of 4 litres and 3.5 litres are also available, but mainly in public buildings. While these appliances are available, it is difficult for the consumer to be able to identify them from available literature.

W.c. replacement cycles are estimated at 15 years; however this can be extended with proper maintenance. Because w.c.'s tend to be replaced for reasons of style or colour, there is a limited window of opportunity to convince consumers of the need to reduce water consumption. Since there is little difference in component design between a 6 litre w.c. and a 4.5 litre w.c., with the exception of smaller cistern size and potentially different bowl shape, cost should not be a significant factor in the replacement of different flush volume w.c.'s (BNWAT20 2007). In 2007 prices for w.c.'s were approximately £163 for a 6/4 litre dual flush, £275 for a 4.5 litre toilet, but only £120 for a 4.5/3 litre dual flush (Environment Agency 2007). While there is not a significant premium for low flush volume w.c.'s, it is recognised that they are unlikely to be significantly cheaper than conventional w.c.'s (BNWAT20 2006).

As a result of the slow replacement cycle of toilets, it will take time to show significant reduction in overall water consumption from this measure alone. It may therefore be worthwhile considering the use of incentives to achieve water reduction goals more quickly. Incentives are used in many countries to encourage users to upgrade to a toilet with a lower flush volume. Most incentives revolve around some monetary reduction in water bills or a rebate for introducing the water saving technology (Canada, USA). While this may pose particular problems in Scotland due to the nature of the water industry and the absence of a link between water usage and water payment, some method of incentive could be considered further. Suggestions include a scrappage scheme, but as there is no evidence of such an incentive scheme in existence it is an idea which could be explored more fully.

Scottish Water (Gormley 2011) is involved in a number of schemes to address this issue. The 'incentives for developers scheme' has been introduced in an attempt to raise awareness of developers on the benefits of water efficiency measures. Plans are also progressing on a water efficiency strategy for consumers and supplier alike. These schemes should be encouraged and made as widely available as possible to encourage the uptake of low flush volume w.c.'s.

7.5.3 Assessment of impact on drainage system performance

A measure of the effectiveness of any particular flush volume is the maximum distance that the flush can transport solids. There is a distance beyond which further flushes will have no effect and solids will be permanently deposited. This maximum distance is due to the attenuation of the surge wave discharged by the w.c. and is characteristic of the w.c. design, the solids discharged and the configuration of the drain. It is therefore the single most useful indicator of system performance under a given set of conditions and is the most suitable comparator of the impact of flush volume on building drain operation.

Two different scenarios were tested by the simulations in this research. The first involves establishing the maximum possible distance that a particular w.c. and drain combination can achieve, thus establishing relative performance. The second scenario tested was to simulate a full house operation over a period of three hours using w.c.'s of different flush volumes and drains of different configurations. This is to assess the likelihood of blockages occurring in a single storey house under normal system usage, the transport of solids being assisted by the combining of flows from additional appliances such as sinks, showers, baths and washing machines.

7.5.4 Scenario one: maximum travel distance

Dual flush volume w.c.'s present a difficulty when running simulations. When in use, the larger of the two volumes will be used to remove waste material while the smaller may only need to transport toilet paper and clean the bowl. The ability of the flush volume to clean the bowl is dealt with in the testing regime for w.c. performance contained in the water byelaws (Water Byelaws 2000), and any w.c. approved for sale must first pass bowl cleaning tests.

The most widely available dual flush w.c.'s on the market are 6 litre/4 litre and 4 litre/2.6 litre, the former being the most common in use at present. The introduction of an 'average' or 'equivalent' flush presents additional issues with regard to the actual volume of flush to be used in simulation runs. The classification of a w.c. by its average flush volume is based on the likely occurrence of each flush over a period of time. Recent research (Environment Agency 2007) has established that a ratio of four of the smaller flush volume flushes to one large flush volume flush is likely in practice. The methodology adopted here to deal with dual flush w.c.'s was to first flush a solid into the drain using the larger of the flush volumes. This solid was then transported by four smaller flush volumes and then a large flush volume until the solid eventually came to rest. This methodology more accurately reflects usage patterns in practice and offers the opportunity to observe the limited trans-portation ability of the smaller flush volumes.

To establish the maximum solid travel distance for a particular flush vol-ume and system configuration, a solid is discharged into a drain from a w.c. and the system is repeatedly flushed. At some point along the pipe the solid will come to rest and be deposited. At this point no further flushes will insti-gate movement (since the surge wave from the w.c. will have attenuated to a depth and velocity below the movement threshold for that solid). This was first established by Swaffield (Swaffield and Galowin 1992) and forms the basis for much of the work in this area as described in Chapter 3, 4 and 5, including the methodology adopted by WRc in CP 367 (Drinkwater 2009).

All simulations described here used solids with the same characteristics. A solid based on a 38 mm diameter cylinder of 80 mm length and a specific gravity of 1.2 was specified. This 'test solid' was first used by the National Bureau of Standards (NBS) in the USA in the 1980s and has formed the basis for much of the testing on innovation in w.c. design and drainline solid trans-port prediction since (Swaffield and Galowin 1992). This solid represents a flush load which is a combination of solids and toilet paper and, since it does not disintegrate, presents the system with a worst case scenario in terms of solid transportability.

Figure 7.10 represents a simple simulation input for scenario one described above to establish maximum travel distance. The w.c. is connected to a long horizontal pipe run. The diameter and the gradient of the pipe run can be changed together with the flush volume of the w.c. in order to establish the maximum travel distance for each configuration.

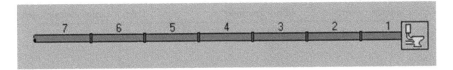

Figure 7.10 Simple installation to test maximum travel distance.

The results for maximum travel distance achievable for the 6 litre/4 litre w.c. are shown in Table 7.1 for various pipe configurations. It can clearly be seen that the best performance is achieved when a pipe diameter of 75 mm is used and set to a gradient of 1 in 40. While it is understood that a pipe diameter of 75 mm is not allowed due to the requirement that a decrease in pipe diameter is not allowed in the direction of flow, results for this configuration are included here, however, to show the superior solid transport properties of this pipe diameter. Possible future innovation in w.c. design may make the introduction of 75 mm w.c. connection inevitable.

In the absence of the possibility of utilising a 75 mm w.c. connecting pipe, the best pipe/gradient configuration is the 100 mm pipe set to a gradient of 1 in 40. This configuration gives a maximum solid transport distance of 56 m which is far in excess of the generally accepted average distance from house to the main sewer connection of 10 m. This maximum distance means that a 6 litre/4 litre flush volume is not likely to cause blockage problems in well maintained drains using this pipe diameter–gradient combination.

Since this flush volume is currently the industry standard it is not surprising that there isn't a problem with this combination; however it does act as a useful baseline for measuring the impact of reducing flush volume further.

Care should be taken on long horizontal runs where no joining flows are possible and a steeper gradient than 1 in 100 is not achievable; then some limits on usage (either distance to the sewer or allowable w.c. installation) should apply.

Maximum transport distance for 4 litre/2.6 litre flush volume

The results for maximum travel distance achievable for the 4 litre/2.6 litre w.c. are shown in Table 7.2 for various pipe configurations. Again it can be seen that the 75 mm connecting pipe set to a gradient of 1 in 40 achieves the best result. While there is a reduction in the maximum possible transport distances for solids overall, the 100 mm pipe diameter set to a gradient of 1 in 40 still achieves an appreciable 40 m transport from the w.c. discharge. This still represents a considerable margin of safety to the notional 10 m average distance to the main sewer described above. There is therefore a level of confidence that this configuration would not contribute to an increased risk of blockage in a free running drain.

Table 7.1 Summary of simulation results for 6 litre/4 litre flush volume w.c.

Pipe diameter (mm)	Pipe gradient	Max travel distance (m)
100	1/40	56
100	1/60	44
100	1/80	28
100	1/100	19
75	1/40	125
75	1/60	75
75	1/80	47
75	1/100	18

Table 7.2 Summary of simulation results for 4 litre/2.6 litre flush volume

Pipe diameter	Pipe gradient	Max travel distance (m)
100	1/40	40
100	1/60	25
100	1/80	15
100	1/100	10
75	1/40	82
75	1/60	45
75	1/80	30
75	1/100	55

There is however a higher risk at shallower gradients – 100 mm pipes set to a gradient of 1 in 100 barely clear the 10 m mark, and care should therefore be taken on long horizontal runs where the likelihood of blockage is greatest. In the extreme case where a single w.c. is connected via a long pipe (greater than 10 m) and there are no joining flows from other appliances and a gradient steeper than 1 in 100 is not achievable, then the use of low flush volume w.c.'s is not recommended.

7.5.5 Scenario two: simulation of real system operation

In order to test the applicability of the recommended 'best practice' drain configurations established from the simulations on maximum travel distance given above, a simulation of the system shown in Figure 7.11 was carried out.

The methodology adopted was to simulate the operation of the installation over a 3 hour morning peak period. To test the applicability to a worst case scenario single occupancy is assumed. The usage patterns incorporated into the simulation were adapted from published surveys (Friedler *et al.* 1996; Gormley and Dickenson 2008). Table 7.3 shows the assumed appliance water usage, and Table 7.4 shows the assumed usage pattern, including time of use and frequency of use during the 3 hour period. This information was programmed into the model and the simulation was run.

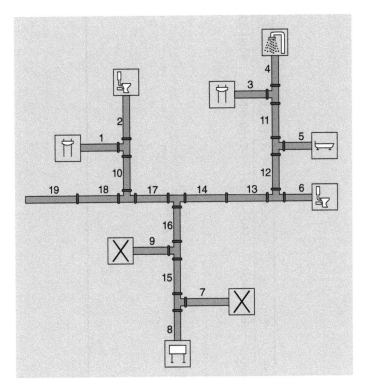

Figure 7.11 Domestic installation used to test the applicability of 4 litre/2.6 litre
w.c. installation – note: representation only, not to scale.

Table 7.3 Appliances and associate volume of water used

Appliance	Water used (litres)	Appliance	Water used (litres)
Bath	75 to 90	Sink	6 to 8
Shower	5 to 7 per min	Wash face & hands	3 to 9
W.c.	4 / 2.6	Dishwasher	15
Washing machine	50		

Source: Adapted from data by 'The Water School'.

The simulation was run for the 3 hour period for the single occupancy
house to assess whether solids discharged into the drain from the furthest
w.c. from the main sewer, represented on Figure 7.11 as the pipe leaving
the house, Pipe 19, would enter the main sewer before the end of the 3 hour
period.

Water entering the drain from the other appliances contributes to the
cleansing of the drain, and it was found that no solids remained in the

Table 7.4 Appliance usage data for house simulation

Appliance	Morn uses	6:00–6:30	6:30–7:00	7:00–7:30	7:30–8:00	8:00–8:30	8:30–9:00
Time (secs)		0–1800	1800–3600	3600–5400	5400–7200	7200–9000	9000–10800
Bath	0						
Shower	1				1		
W.c. 1 (full flush)	1	1					
Basin 1	1	1					
W.c. 2 (half flush)	1					1	
Basin 2	1		1			1	
Washing machine	1		1				
Dishwasher	1						
Sink	1			1			

Source: Adapted from Gormley and Dickenson (2008).

system after the 3 hour period for 100 mm pipe diameter set to slopes of 1 in 40, 1 in 60, 1 in 80 or 1 in 100. The poor maximum travel distances achieved by the shallower gradients in the previous simulations were compensated for by the joining flows from other appliances. It should be noted that since a shower was taken no bath discharge was incorporated into the simulation.

7.5.6 Code validation: BS EN codes

BS EN 12056: 2000 allows for all the configurations described above and provides the design guidance necessary to ensure proper system operation when the w.c. flush volume is reduced to 4 litres full flush and 2.6 litres half flush or 3.5 litre average flush, as long as the drain is in a good state of repair and the system is not unusually overloaded. While BS EN 12056: 2000 permits the use of pipe diameters and gradients tested in this research, there are no restrictions on pipe length given. The results shown in Table 7.1 and Table 7.2 suggest that there should be a maximum length of pipe in the case where no joining flows are possible.

EN 752:2008 covers the design and maintenance of drain and sewer systems outside buildings and is still considered suitable for purpose in view of the reductions in water volume discharged from lower flush volume w.c.'s. A 2 litre per flush reduction in flush volume reduces the overall water loading on a drain/sewer by only a small fraction, and it is fully anticipated that the risk of blockage is low as long as the sewer is not unusually overloaded or the sewer line has no other contributing flows to maintain self cleansing

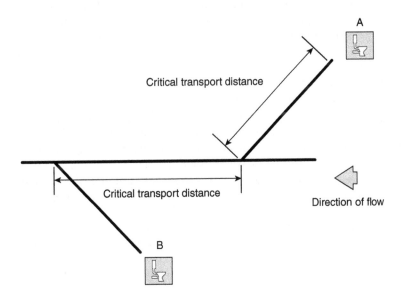

Figure 7.12 Importance of adjoining flows.

velocity. EN752:2008 allows for steeper slopes to be installed in cases where increased cleansing velocity is required; however care must be taken with entrained airflows in the sewer as a result.

The simulations have confirmed that the risk of blockage associated with introducing properly installed 4 litre/2.6 litre flush volume w.c.'s into a properly maintained and defect free building drainage system is relatively low. The risk of blockage increases with shallower gradients such as 1 in 100, and significantly better performance is obtained from gradients of 1 in 40. Many older installations may have their drainage pipes installed at a gradient of 1 in 40; it is assumed therefore that such systems would not be unduly compromised from the introduction of lower flush volume w.c.'s. The contribution made to drain system self cleansing by joining flows for other appliances is significant and should be taken into account when planning an installation. For example, a long horizontal pipe run from an individual w.c. set to a shallow gradient with no joining flows from other appliances should be avoided. To assess the gradients that drainage systems have generally been installed to, consideration should be given to a survey of the information held by local authority verifiers.

7.5.7 A note on adjoining flows

In both the case studies described above there is an underlying concern relating to the amount of water available to maintain free running drains. The root of many concerns from others regarding pressures to reduce water consumption further is the idea that, for a given w.c. in a given installation, there is a maximum travel distance possible. This is of concern because it represents a clear and present risk of blockages occurring. One aspect of an installation, the design of which does not feature in any design codes, is the distance to adjoining flow. The situation is illustrated in Figure 7.12.

The w.c.'s labelled A and B could just as easily be labelled 'House A' and 'House B' – they represent an independent system or an independent branch on a system connecting to a collection drain. The figure shows that there are two critical transport distances: one from appliance A to the collection drain and the second from the joining point from Appliance A to the joining point from Appliance B. Under low water usage criteria these distances are critical to ensure there are no blockages in the system. The amount of water required still depends on the type of appliance being used and the usage patterns. The arrangement shown here can only really be assessed using a modelling method such as the one described here; discharge units or drainage loading or fixture units give an indication of probable loading but do not assist in the assessment of critical transport distances.

7.6 Concluding remarks

The example applications shown in this chapter highlight the benefits of modelling free surface water flows and solid transport in building drainage systems. The speed with which even the more involved scenarios could be simulated (a matter of minutes run time) means that small adjustments can be made to scenarios and flow assumptions in order to exhaust all possible outcomes. While the Method of Characteristics model gives little detail in a three-dimensional sense of the flows at different stages of development through a system, it greatly enhances outcomes to have fast calculation times for local flow rate, velocity and depth along the entire length of pipework in a network. By contrast, other prediction methods such as computational fluid dynamics (CFD) offer greater localised detail in all three dimensions and at all points throughout the network; however complex system geometry setup and long calculation times still preclude its use as a fast prediction method. This chapter confirms the applicability of the Method of Characteristics–based modelling technique for free surface flows as a suitable method with which to influence legislature and coding bodies alike.

8   Afterword

This book has sought to set out the fundamental engineering science relating to the prediction and simulation of building drainage free surface flows, based on nearly 40 years of research and investigation. This has led to a comprehensive approach to designing building drainage systems, and understanding the '*why*' as well as the '*how*' of system design. The fundamental mathematics for the description of free surface waves has been available since the nineteenth century when the French elasticity theorist Adhémar Jean Claude Barré de Saint-Venant, commonly known as St Venant, developed his 1-D equations to model open channel flow. These 1-D equations are easier to solve than the much more complex 'shallow water equations' and have been shown (see Chapter 3) to be wholly appropriate for free surface wave modelling in partially filled building drainage pipes (Waterwise 2014).

While these equations have been around for a long time it has not been until the advent of the digital computer that modelling to any great extent has been carried out. This 'lag' between the time when design methodologies were needed and the time when high-speed predictive modelling became available meant that alternative, simpler methods had to be adopted to allow the sanitary engineer to design a system as safely and as efficiently as possible. This task is made more difficult for many public health engineering practitioners, since an accepted methodology advocated by codes and standards worldwide is the 'discharge unit' or the 'discharge fixture unit' method which is based on the steady state probabilistic discharge of fixtures into a drain, as described in Chapter 2. A summation of these discharge units allows the designer to select an appropriate pipe diameter and gradient. While this system offers simplicity and ease of pipe diameter selection, it is quite rigid and allows the designer little scope for manoeuvre – a system of design, simulation and assessment would provide much greater insight into the operation of the system under a range of different usage scenarios, thus making the design more appropriate in an industry and field of engineering constantly changing and updating.

Much of current design methodology is based on the work of Dr. Roy Hunter, who worked for the National Bureau of Standards in the USA for over two decades. His significant publication 'Methods of estimating loads on plumbing systems, BMS 65' (Hunter, 1940) appeared in 1940 with the noted purpose:

> [The] Purpose of this series of papers is to collect in organized manner information obtained by the author over a number of years (from 1921 to research of 1937–1940) on plumbing, together with results from research (1937–1940) on plumbing with intervening experiments and interpreting results in a form suitable for direct and practical applications. It is hoped that this series of papers will supply logical answers to questions pertaining to pipe sizes and design of plumbing construction.

Hunter's work on dimensionless discharge and fixture units for plumbing systems was seminal and introduced binomial probability theory applied to system demands for the first time. His intention was to describe the detailed calculations and experiments upon which the probabilistic units were based in his 1940s series. Unfortunately, Hunter died in 1943 leaving only the design system linking discharges to pipe sizing, with the basis for the calculations sadly absent. Modifications have been made over the years to account for water conservation strategies; however the basic design method is still widely used, despite an awareness that numerical methods may be more beneficial (Galowin, 2009). That this is still a well-used design methodology plays a significant role in the rationale for this book. The following discussion sets out what is required in order to fully understand and predict free surface flows relating to building drainage systems post Hunter.

The fluid mechanics described in Chapter 2 is intended to form the basis for water flow fundamental equations and their derivations, placing the work on building drainage systems in a wider context. Waste-water flows in building drainage systems can be easily categorised as free surface flows in partially filled pipes (with the odd exception of full bore flow – see Chapter 6). The flows themselves are generated by random discharges from appliances thus complicating the issue.

The development of the 1-D St Venant equations discussed above forms the mathematical basis for wave propagation in a system. Central to the interpretive power of such models is the ability to predict and model wave attenuation along a pipe. This fundamental opens up opportunities for assessing surge waves generated by discharging appliances, which in itself lies at the core of any analysis. An insight into how the fundamental equations are transformed and made ready are discussed in Chapter 3 for the application of the method to the building drainage system in Chapter 4. Boundary condition derivation and validation play a key role in allowing the calculations to progress and are central to the application of the method.

While predicting waste water flows in building drainage systems is important, it should be remembered that the function of the water is to carry human waste away from habitable space. This is the fundamental purpose of the system, and it could be argued, it is the job of the engineer/researcher to make sure that this function occurs as efficiently as possible. Since approximately one-third of the water in the drain originates from the w.c., it must be remembered that any water-conserving initiative is going to influence how much water is available to clean the drain and remove solids, mainly faeces and toilet paper. The moving boundary condition associated with solid transport has proved a challenge over the years. The velocity decrement method (Swaffield and McDougall 1996) and the mach number model method (Gormley and Campbell 2005, 2006) as applied to interacting solids are both discussed in Chapter 5.

The application of flow modelling to rainwater drainage systems, both conventional and siphonic, has much to do with 'too much water'. Chapter 6 describes the underlying principles which are illustrated through application of some of the many numerical models that have been developed over the past 20 years at Heriot-Watt University (FM4GUTT, SIPHONET, ROOFNET).

Of the two basic types of rainwater drainage system, conventional systems are the simplest to model. They operate at atmospheric pressure, and the driving head is thus limited to the gutter flow depth, whereas that within siphonic systems is designed to run full bore, resulting in sub-atmospheric system pressures, higher driving heads and higher flow velocities. As a result, siphonic systems normally require far fewer downpipes to drain the same size roof area, and the depressurised conditions also mean that much of the collection pipework can be routed at high level, thus reducing the extent of costly underground pipework. These advantages make siphonic systems ideally suited to large industrial and commercial roof areas, with high profile examples including the Sydney Olympic stadium and Hong Kong Airport, as well as historically Stansted Airport, Edinburgh Murrayfield Stadium and many other commercial applications. The ability to fully describe and predict pressures and water flows in siphonic systems requires more than the 1-D St Venant equations. Additional techniques such as the MacCormack method developed in Chapter 6 can be employed to simulate initial free surface flow conditions, whilst retaining the MoC technique for the simulation of full bore flow conditions. The transition between the two flow regimes can be dealt with using the Preissmann slot technique developed in Chapter 3. Combined, these methods produce a suite of numerical models capable of accurately simulating conditions within all types of rainwater drainage systems, under a variety of different loading regimes.

The derivation of equations and the development of boundary conditions is an excellent output from a research project, but it does not show that these methods have real impact in the industry and with legislative bodies. In Chapter 7 the model is applied to two real-life questions facing building control regulators; in the first, a question is posed regarding the likely impact of reducing w.c. flush volumes even lower than those currently legislated for, 6 litres maximum. In the second case study, the impact of increasing the diameter of the connection pipe between a house and the main sewer is assessed. In both cases it is possible to use the predictive models developed throughout this book to make an informed opinion, based on quantitative data with associated qualitative assessment to guide legislators – without having to construct costly test rigs or initiate pilot studies.

It cannot be denied that we live in a rapidly changing world in which we have the conflicting challenges of 'too much water' and 'too little water'. The difficulty in providing potable water in those areas where water stress is a reality is a challenge for engineers, designers, manufacturers, code authorities and legislators alike, and paradoxically, the same group of people must also deal with the conflicting challenge of dealing with the disposal of excess water from increased intensity precipitation.

This is no time for comfortably reassuring ourselves that we can adequately deal with the relentless march of progress with the application of old steady-state methodologies. It is now a necessity that predictive methods become mainstream. The work carried out over the last 40 years, first at Brunel University and latterly at Heriot-Watt University sets the basis for a complete methodology with which to accurately predict how these expected stresses will affect the building drainage system performance, regardless of the nature of the change itself; be it water conservation initiatives requiring confirmation that lower w.c. flush volumes are adequate, as described in Chapter 7, or the changing nature of rainwater management, as described in Chapter 6, modelling and simulation is the only viable option for the validation of adaptation techniques.

It is hoped that the reader will find this treatise on free surface wave modelling a logical discussion on the methods and techniques required in order to successfully model flows in building drainage systems. The extensive use of the developed models, namely DRAINET and ROOFNET, SIPHONET and FM4GUTT to illustrate the validity of solutions to the real engineering problems and hypotheses presented, is in itself testimony to their usefulness. Where possible, validation by laboratory or site investigation has been used to corroborate model outputs.

This book represents almost 40 years of research and effort by a considerable number of people. Its main aim is to highlight the applicability of engineered design to a field of engineering hitherto exposed only to design guidance based on tabulated flows and pipe sizes. These methods do have a mathematical basis, but they don't tell the whole story, and the whole story is now, more than ever before, required in order to equip the engineer with the appropriate tools with which to tackle the relentless problems caused by a rapidly changing world.

References

Abbott, M.B. and Minns, A.W. (1998) *Computational hydraulics*. Aldershot: Ashgate.

American Society of Mechanical Engineers (ASME) (2003) 'Vitreous china plumbing fixtures and hydraulic requirements for water closets'. ASME A112.19.2-2003.

American Water Works Association Research Foundation (1999) 'Water efficiency making cents in the next century'. *AWWA Proceedings*, 'Conserv 99' Monterey, California, 31 January–3 February.

Arnell, N.W. (2004) *'Climate change and global water resources: SRES emissions and socio-economic scenarios'*. Global Environmental Change 14 (2004): 31–52.

Arthur, S. and Swaffield, J.A. (2001) 'Siphonic roof drainage system analysis utilising unsteady flow theory'. *Building and Environment* 36, 8: 939–948.

Arthur, S., Wright, G. and Swaffield, J. (2005) 'Operational performance of siphonic roof drainage systems'. *Building and Environment* 40, 6: 788–796.

Beecham, S., Baddoo, N., Hall, T., Robinson, H. and Sharp, H. (2008) 'Motivation in software engineering: a systematic literature review'. *Information and Software Technology* 50, No. 9–10 (August 2008): 860–878.

Benn, H. (2009) '"UK rainfall scenarios", Launch of UK government predictions'. *Guardian*, 19 June.

Billington, N.S. and Roberts, B.M. (1982) *Building services engineering: a review of its development*. Pergamon Press.

BMA (Bathroom Manufacturers Association) (2010) *Bathroom Trends*.

BNWAT20 (2007) Very low water use water closets – Innovation Briefing Note. Market Transformation Programme.

Bocarro, R.A. (1987) 'Water conservation w.c. design for developing countries'. PhD thesis, Brunel University, UK.

Bokor, S.D. (1982) 'Correlation of laboratory and installed drainage system solid transport'. PhD thesis, Brunel University, Uxbridge.

Bridge, S. (1984) 'A study of unsteady flow wave attenuation in partially filled pipe networks'. PhD thesis, Brunel University, UK.

Bridge, S. and Swaffield, J.A. (1983) 'Applicability of the Colebrook-White formula to represent frictional losses in partially filled unsteady pipeflow'. *Journal of Research, NBS*, Vol. 88 No. 6, Nov–Dec: 389–393.

BSI (2000) BS EN 12056-3:2000 *Gravity drainage systems inside buildings. Roof drainage, layout and calculation*. UK: British Standards Publishing Limited.

Burberry, P.J. (1978) 'Water economy and the hydraulic design of underground drainage'. CIB W62 Seminar, Brussels.

Butler, D. and Davies, J.W. (2004) *Urban drainage*. 2nd Edition. London: Spon.

Butler, D., Davies, J.W., Jeffries, C. and Schutze, M. (2003) 'Gross solid transport in sewers'. *Procs. ICE, Water and Maritime Eng.* 156, June, Issue WM2: 175–183.

Chartered Institution of Building Services Engineers (2009) *CIBSE Guide G, Public Health Engineering.* CIBSE Publications.

Chow, V.T., Maidment, D.R. and Mays, L.W. (1988) *Applied Hydrology.* Singapore: McGraw-Hill.

Clark, M. (1994) 'The measurement of solid velocities in building drainage systems'. Msc thesis, Heriot-Watt University, Scotland.

Cox, D. (1997) 'Designing with water'. *The Society for Responsible Design Newsletter* 48: 4–16.

Crerar, S.A. (1989) 'A study of the transport of discrete solids in building drainage systems'. PhD thesis, Heriot-Watt University, Scotland.

Cummings, S. (2003) 'The design of water closet systems in Australia with particular reference to minimising water consumption'. Doc. of Env. Design thesis, University of Canberra.

Cummings, S. (2010) 'Outcomes from an industry and regulator research collaboration into reduced flows on building drainline systems in Australia'. CIB W62 Conference Water Supply and Drainage for Buildings, Sydney, Nov. 8–10th: 478–487.

Cummings, S. and de Marco, P. (2010) 'A status report: reduced flows in building drains' Water Smart Innovations. Las Vegas October 6–8.

Cummings, S., McDougall, J.A. and Swaffield, J.A. (2007) 'Hydraulic assessment of noncircular section drains within a building drainage network'. *BR and I* Volume 35, No. 3: 316–328.

Davies, J.W., Butler, D. and Xu, Y.L. (1996) 'Gross solid movement in sewers: laboratory studies as a basis for a model'. *Jour. Institution of Water and Env. Mangt.* 10(1): 52–58.

Davies, J.W., Schluter, W., Jeffries, C. and Butler, D. (2002) 'Laboratory and field studies to support a model of gross solids transport'. *Global Solutions for Urban Drainage*, 9th Int Conf on Urban Drainage, Portland Oregon, 8–13.

de Marco. P. (2009) Personal communication – objectives of the PERC programme.

DCLG (2009) Demographic predictions drawn from the current Housing Statistical Release 11 March 2009 DCLG. http://www.statistics.gov.uk/statbase/Product. asp?vlnk=997.

Department of the Environment, Food and Rural Affairs (Defra) (1999) 'Water Supply (Water Fittings) Regulations 1999'. HMSO London, July.

Diaper, C., Jefferson, B., Parsons, S.A. and Judd, S.J. (2001) 'Water-recycling technologies in the UK'. *Journal of CIWEM*, 15: 282–286.

Dixon, A., Butler, D. and Fewkes, A. (2000) 'Influence of scale in grey water reuse systems', as cited by Jefferson, B., Laine, A., Parsons, S., Stephenson, T. and Judd, S. (1999) 'Technologies for domestic wastewater recycling'. *Urban Water*, 1(4): 285–292.

Dolnicar, S. and Schafer, A.L. (2009) 'Desalinated versus recycled water: public perceptions and profiles of the acceptors'. *Journal of Environmental Management*, 90 (2): 888–900.

Douglas, J.F., Gasiorek, J.M., Swaffield, J.A. and Jack, L.B. (2005) *Fluid mechanics*, 5th edn. Pearson.

Drinkwater, A. (2009) 'Pull the chain, fill the drain'. CP 367 – The effect of reduced water usage on sewer solid movement in small pipes Report no.: 7904, Water Research Council (WRc).

Drinkwater, A., Moy, F. and Poinel, L. (2009) 'CP 367 – the effect of reduced water usage on sewer solid movement in small pipes', Collaborative Project CP: 367 WRc, Report No P8086, October.

Edwards, K. and Martin, L. (1995) 'A methodology for surveying domestic water consumption'. *Journal CIWEM*, 9 October: 177–188.

EN752:2008 'Drain and sewer system outside buildings'. London: British Standards Institute.

EN 12056: 2000 'Gravity drainage systems inside buildings part 2: Sanitary pipework, layout and calculations'. London: British Standards Institute.

Environment Agency (2005) 'Water Savings Group promises action'. Water Demand Management Bulletin, Issue 74, December.

Environment Agency (2007). 'Assessing the cost of compliance with the code for sustainable homes.

Environment Committee (1996) 'Water conservation and supply', First Report, Session 1996–97. London: The Stationary Office.

Evans, E., Ashley, R., Hall, J., Penning-Rowsell, E., Saul, A., Sayers, P., Thorne, C. and Watkinson, A. (2004) *Foresight. Future Flooding. Scientific Summary: Volume I – Future risks and their drivers*. London: Office of Science and Technology.

Ezekial, F.D. and Paynter, H.M. (1957) 'Computer representation of engineering systems involving fluid transients'. *Trans. ASME*, Vol. 79, Paper No. 56-A-120: 1840–1850.

Fennema, R.J. and Chaudhry, M.H (1989) 'Implicit methods for two dimensional unsteady free surface flows'. *Journ. Hydr. Res. Interational Association of Hydraulic Research*, 27(3): 321–332.

Fox, J.A. (1968) 'The use of digital computers in the solution of waterhammer problems'. *Proc. ICE*, Vol. 39: 127–131.

French, R.H. (1985). *Open channel hydraulics*. New York: McGraw Hill.

Friedler, E., Bulter, D. and Brown, D.M. (1996) 'Domestic WC usage patterns'. *Building and Environment*, Vol. 31 No. 4: 385–392.

Galowin, L.S. (1979) 'Review of HUD/NBS Residential Water Conservation'. Conseil International du Batiment CIB W62 Symposium, CSTB Paris, 1979, November.

Galowin, L.S. (1982) 'Sweeping of solids in drains with steady flows'. CIB W62 Symposium Water Supply and Drainage for Buildings, Lostorf, Switzerland, 1982, September.

Goldberg, D.E. and Wylie, E.B. (1983) 'Characteristics method using time-line interpolations'. *J. Hyd. Div. ASCE*, Vol. 109, No.5: 670–683.

Gormley, M. (2005) 'After the tsunami: water supply and sanitation, from emergency response to rehabilitation.' *CIB W62 International Symposium on Water Supply and Drainage for Buildings,* Brussels, September.

Gormley, M. (2011) 'Assessment of the impact of low flush volume WCs on building drainage system operation'. Scottish Government Report.

Gormley, M. and Campbell, D.P. (2006a) 'Modelling water reduction effects: method and implications for horizontal drainage.' *Building Research & Information* 34(2) (2006): 131–144.

Gormley, M. and Campbell, D.P. (2006b) 'The transport of discrete solids in above ground near horizontal drainage pipes: A wave speed dependent model'. *Building and Environment* 41 (2006): 534–547.

Gormley, M. and Campbell, D.P. (2007) 'The effects of surfactant dosed water on solid transport in above ground near horizontal drainage systems'. *Building and Environment* 42: 707–716.

Gormley, M. and Dickenson, S.K. (2008) 'Implications of greywater reuse on solid transport in building drainage systems'. CIB W62 International Symposium on Water Supply and Drainage for Buildings, Hong Kong.

Gormley, M., Mara, D.D., Jean, N.J. and McDougall, J.A. (2013) 'Pro-poor sewerage: solids modelling for design optimization'. *Proceedings of the Institution of Civil Engineers (ICE): Municipal Engineer.*

Gray, C.A.M. (1953) 'Analysis of the dissipation of energy in waterhammer'. *Procs. ASCE*, Vol. 119, Paper 274: 1176–1194.

Griggs, J. and Shouler, M. (1994) 'An examination of water conservation measures'. Proceedings CIB W62 Symposium on Water Supply and Drainage for Buildings, Brighton 26–30.

Hills, S., Birks, R., and McKenzie, B. (2002) 'The Millennium Dome "Watercycle" experiment: to evaluate water efficiency and customer perception at a recycling scheme for 6 million visitors'. *Water, Science and Technology,* 46(6–7): 233–240.

Howarth, G., Swaffield, J.A. and Wakelin, R.H.M. (1980) 'Development of a flushability criterion for sanitary products'. Conseil International du Batiment CIB W62 Symposium, Brunel University July 1980.

Hunter, R.B. (1924) 'Minimum requirements for plumbing', US Department of Commerce, 1924 Plumbing Code.

Hunter, R.B. (1940) 'Methods of estimating loads on plumbing systems', BMS 65 and BMS 79, Washington, DC National Bureau of Standards.

Institute of Hydrology (2005) Flood estimation handbook (FEH).

Jefferson, B., Laine, A., Parsons, S., Stephenson, T. and Judd, S. (2000) 'Technologies for domestic wastewater recycling'. *Urban Water,* 1(4): 285–292.

Jolly, J.P. and Yevjevich, V. (1971) 'Amplification criterion of gradually varied, single peaked waves'. *Hydrology* Paper 51. Colorado State University, Fort Collins, Colorado.

Kamata, M., Matsuo, Y. and Tsukagoshi, N. (1979) 'Studies on flow and transport of faeces in horizontal waste pipes'. CIB W62 Symposium, CSTB Paris, November.

Lamoen, J. (1947) 'Le coup de belier d'Allievi, compte tenu des pertes de charge continues'. Bull. Centre de Etudes, de Recherches et d'Essais Scientifiques des constructions du Gerrie Civil et Hydraulique Fluviale, Tome II, Doscor, Liege.

Lauchlan, C., Griggs, J. and Escarameia, M. (2004) 'Drainage design for buildings with reduced water use'. BRE publication IP 1/04, Watford.

Lazarova, V. (2001) 'Role of water reuse in enhancing integrated water management in Europe'. *Final Report of the EU project CatchWater,* V. Lazarova (Ed.), ONDEO, Paris, France, p. 708, as cited in Lazarova, V., Hills, S. and Birks, R. (2003) 'Using recycled water for non-potable, urban uses: a review with particular reference to toilet flushing'. *Water, Science and Technology: Water Supply,* 3 (4): 69–77.

Lillywhite, M.S.T and Webster, C.J.D. (1979) 'Investigations of drain blockages and their implications of design'. *Journal of Institution of Public Health Engineers,* Vol. 7 No. 2: 53–60.

Lister, M. (1960) 'Numerical simulation of hyperbolic partial differential equations by the method of characteristics', in Ralston and Wilf (eds.) *Mathematical Methods for Digital Computers.* New York: John Wiley, pp. 165–179.

Littlewood, K. (2000) 'Movement of gross solids in small diameter sewers'. Unpublished PhD thesis, Imperial College, London.

Littlewood, K. and Butler, D. (2003) 'Movement mechanisms of gross solids in intermittent flow'. *Water Science and Technology,* Vol. 47: 45–50.

Mahajan, B.M. (1981a) 'Unsteady water depth measurement in a partially filled 7.6 cm diameter horizontal pipe'. NBSIR 81-2249, April.

Mahajan, B.M. (1981b) 'Experimental investigation of transport of finite solids in a 76 mm diameter partially filled pipe'. NBSIR 81-2266, April.

Marriott, B.S.T. (1978) 'Transport of solids with reduced flush volumes'. Unpublished dissertation, Brunel University, Uxbridge, 1978.

Maxwell Standing, K. (1986) 'Improvements in the application of the numerical method of characteristics to predict attenuation in unsteady partially filled pipe-flow'. *Jour. of Research, NBS*, 91 (3), May–June.

May, R.W.P. and Escarameia, M. (1996) Performance of siphonic drainage systems for roof gutters. Report No. SR463, Wallingford, England: HR Wallingford Limited.

McDougall, J.A. (1995) 'Mathematical modelling of solid transport in defective building drainage systems'. PhD thesis, Heriot-Watt University, Edinburgh.

McDougall, J.A. and Swaffield, J.A. (2000) 'Simulation of building drainage system operation under water conservation design criteria'. *BSER&T*, Vol. 21, No. 1: 41–52.

McDougall, J.A. and Swaffield, J.A. (2003) 'The influence of water conservation on drain sizing for building drainage systems'. *Building Services Engineering Research and Technology*, Vol. 24, No. 4: 229–244.

Moran, D., MacLeod, M., McVittie, A., Lago, M. and Oglethorpe, D. (2007) 'Dynamics of water use in Scotland, Water'. *Water and Environment Journal* Vol. 21: 241–251.

Mozayeng, B. and Song, C.S. (1969) 'Propagation of flood waves in open channels'. *J. Hyd. Div., A.S.C.E.*, 95: HY3.

Murphy, J.M., Sexton, D.M.H., Jenkins, G.J., Boorman, P.M., Booth, B.B.B., Brown, C.C., Clark, R.T., Collins, M., Harris, G.R., Kendon, E.J., Betts, R.A., Brown, S.J., Howard, T.P., Humphrey, K.A., McCarthy, M.P., McDonald, R.E., Stephens, A., Wallace, C., Warren, R., Wilby, R. and Wood, R.A. (2009) *UK Climate Projections Science Report: Climate change projections*. Exeter: Met Office Hadley Centre.

Neilsen, V. (1973) 'Discharge characteristics of sanitary appliances', CIB W62 Symposium Water Supply and Drainage for Buildings, Swedish Building Research Institute, Stockholm, Sweden.

Office of the Deputy Prime Minister (ODPM) (2001) Building Regulations 'Part H', London HMSO.

ODPM (DCLG) (2006) 'Affordability and the Supply of Housing', House of Commons ODPM: Housing, Planning, Local Government and the Regions Committee, Third report of Session 2005-06, HC703-1, May 2006.

Oengoeren, A. and Meier, B. (2007) 'A numerical approach to investigate solid transport characteristics in waste water drainage systems'. CIB W62 International Symposium On Water Supply And Drainage, Brno, Czech Republic.

Orloski, M.J. and Wyly, R.S. (1978) 'Performance criteria and plumbing system design', NBS Technical Note 966 August.

Pender, G., Morvan, H.P., Wright, N.G. and Ervine, D.A., (2005). 'CFD for environmental design and management'. *Computational Fluid Dynamics – Applications in Environmental Hydraulics*. Vol. CH. 18: 487–509.

Pennycook, K., Churcher, D. and Bleicher, D. (2007) *Guide to HVAC building services calculations*. BSRIA.

Rosrud, T. (1977) 'User requirements regarding drainage pipework; development and use of performance specifications', CIB W62 Symposium Water Supply and Drainage for Buildings, Norwegian Building Research Institute, Oslo, May.

Scottish Water (2010) 'Scottish Water Carbon Footprint Report 2008–2009'.

Shaw, E. (2008). *Hydrology in practice.* 3rd Edition. London: Chapman and Hall.

Simms, T., Litlewood, K. and Drinkwater, A. (2006) 'Understanding blockages in small diameter pipes'. WRc report No. P6956.

St. Venant, A.J.C.B. (1870) 'Elementary demonstration of the propagation formula for wave in a prismatic channel'. *Comptes rendus des séances de l'Academie des Sciences,* 71: 186–195.

Streeter, V.I. and Lai, C. (1962) 'Waterhammer analysis including fluid friction'. *Jour. Hyd. Div, ASCE,* Vol. 128, Paper No. 3502, Part 1: 1491–1552.

Surendran, S. and Wheatley, A.D. (1998) 'Grey water reclamation for non-potable reuse'. *Journal of CIWEM,* 12: 406–413.

Swaffield, J.A. (1980a), 'An initial study of the application of the method of characteristics to unsteady flow analysis in partially filled pipe flow'. NBS Report.

Swaffield, J. A. (1980b), 'Application of the method of characteristics to predict attenuation in unsteady partially filled pipe flow'. NBS Report.

Swaffield J.A. (1980c) 'Building drainage system research, past influences, current efforts and future objectives', Institute of Building Construction papers, 1(1): 45–61.

Swaffield, J. A. (1980d), 'Dependence of model waste solid transport characteristics in drainage systems on solid geometry, mass and system parameters'. NBS Report.

Swaffield, J.A. (1981a), 'Calculation techniques for solid friction factors and transport performance incorporating a range of solid buyancy models'. Drainage Research Group, Report No. DreG/NBS/5. Brunel University.

Swaffield, J.A. (1981b) 'The prediction of floating solid velocities in unsteady partially-filled pipe flow.' Drainage Research Group. Brunel University, U.K.

Swaffield, J.A. (1982) 'Application of the method of characteristics to model the transport of discrete solids in partially filled pipe flow'. NBS Building Science Series BSS139, February.

Swaffield, J.A. (1983) 'The prediction of floating solid velocities in unsteady partially filled pipe flow'. NBSIR 83-2614 July, National Bureau of Standards, U.S. Dept of Commerce.

Swaffield, J.A. (2008) 'Modelling low amplitude air pressure transient propagation in building drainage and vent systems to allow system analysis and control'. Proc. of CIBW62 symposium on Water Supply and Drainage for Building, Dusseldorf.

Swaffield, J.A. (2009a) 'Dry drains: myth, reality or impediment to water conservation'. CIB W62 Water Supply and Drainage for Buildings, Dusseldorf, 9–11 September.

Swaffield, J.A. (2009b) 'Inaugural Address of President of The Chartered Institution of Building Services Enginners', London.

Swaffield, J.A. (2010) *Transient airflow in building drainage systems.* Oxen: Spon Press.

Swaffield, J.A. and Boldy, A.P. (1993) *Pressure surge in pipe and duct systems.* Aldershot: Avebury Technical, Gower Press,.

Swaffield, J.A. and Galowin, L.S. (1989) 'Multistorey building drainage network design – an application of computer based unsteady partially filled pipe flow analysis'. *Building and Environment,* Vol. 24, No.1, January: 99–110.

Swaffield, J.A. and Galowin, L.S. (1992) *The engineered design of building drainage systems*. Aldershot: Ashgate, Gower Press.

Swaffield, J.A. and Galowin, L.S. (1993) 'Investigation of the apparent limits of drainline waste transport with low flow volume flush water closets'. Proc. of CIBW62 symposium on Water Supply and Drainage for Buildings, Porto, Portugal, September.

Swaffield, J.A. and Marriott, B.S.T. (1978). 'An investigation of the effect of reduced volume w.c. flush on the transport of solids in above ground drainage systems'. Proc. Of CIBW62 Symposium, Brussels, Belgium.

Swaffield, J.A. and McDougall, J.A. (1996) 'Modelling solid transport in building drainage systems'. *Water Science and Technology*, Vol. 33, No. 9, July.

Swaffield, J.A., McDougall, J.A. and Campbell, D.P. (1999) 'Drainage flow and solid transport in defective building drainage networks'. Proceedings CIBSE Series A: *BSER&T* Vol. 20 No. 2 1999.

Swaffield, J.A. and Thancanamootoo, A. (1991) 'Modelling unsteady annular downflow in vertical building drainage stacks'. *Building and Environment*, Vol. 26, No. 2: 137–142.

Swaffield, J.A. and Wakelin, R.H.M (1976) 'Observation and analysis of the parameters affecting the transport of waste solids in building drainage systems'. *Public Health Engineer, November*.

Swaffield, J.A. and Wakelin, R.H.M. (1996) 'Water conservation: the impact of design, development and site appraisal of a low volume flush toilet'. Chapter 19 in *Low-cost Sewerage*, Mara, D. (Ed.). Chichester: John Wiley and Sons.

Thancanamootoo, A. (1991) 'Unsteady annular downflow in building vertical stacks'. PhD thesis Heriot-Watt University.

Toro, E.F. and Titerev, E.V. (2005) 'ADER schemes for scalar non-linear hyperbolic conservation laws with source terms in 3 space dimensions'. *Journ. Comp. Phys.* 202: 196–215.

Tsukagoshi, N. and Matsuo, Y. (1975) 'Performance of excreta PVA models in drainage systems', CIBW62 Seminar Glasgow University, Scotland.

UK Water Supply (Water Fittings) Regulations 1999.

UN (2011) Global Annual Assessment of Sanitation and Drinking Water (GLAAS).

UNWWAP and UNESCO (2004) Water for People: Water for Life – The UN World Water Development Report.

Uujamhan E.J.S. (1981) 'Water conservation w.c. design: a study of the design parameters affecting w.c. performance'. PhD thesis, Brunel University, U.K.

Vardy, A.E. (1976) 'On the use of the method of characteristics for the solution of unsteady flows in networks', 2nd Int. Conference on Pressure Surge, BHRA, London: 15–30.

Vardy, A.E. (1990), *Fluid principles,* McGraw-Hill, U.K.

Vardy, A.E and Pan, Z. (1998) 'Interpolation in transient polytropic flow'. *Computers & Fluids*, Vol. 27 No.3: 783–796.

Wakelin, R.H.M. (1978) 'A study of the transport of solids in hospital above ground drainage systems'. PhD thesis, Brunel University.

Walski, T., Edwards, B., Helfer, E. and Whitlam, B. (2009) 'Transport of large solids in sewer pipes'. *Water Environment Research*, Vol. 81 No 7: 709–714.

Walski, T., Falco, J., McAloon, M. and Whitman, B. (2011) 'Transport of large solids in unsteady flow in sewers'. *Urban Water Journal, 8(3):* 179–187.

Water Byelaws, 2000 Scotland.

Waterwise (2012) 'Water —the facts'. http://www.waterwise.org.uk/data/resources/25/Water_factsheet_2012.pdf accessed October 23, 2014.

Wise, A.F.E. (1973) 'Research for better sanitary services'. Building Research Establishment, Vol. 95, No. 5.

Wright, G.B. (1997) 'Mathematical modelling of the sub-surface building drainage networks and their associated mechanical vent systems'. PhD thesis, Heriot-Watt University, Scotland.

Wright, G.B., Arthur, S. and Swaffield, J.A. (2006) 'Numerical simulation of the dynamics operation of multi-outlet siphonic roof drainage systems'. *Building and Environment*, 41(9): 1279–1290.

Wyly, R.S. (1964) 'Investigation of hydraulics of horizontal drains in plumbing systems'. Monograoh 86, Washington, DC: National Bureau of Standards.

Yen, B.C. (1986) *Hydraulics of sewers, advances in hydroscience*, Vol. 14. New York: Academic Press.

Index